PACKA
PERSPECTIVES

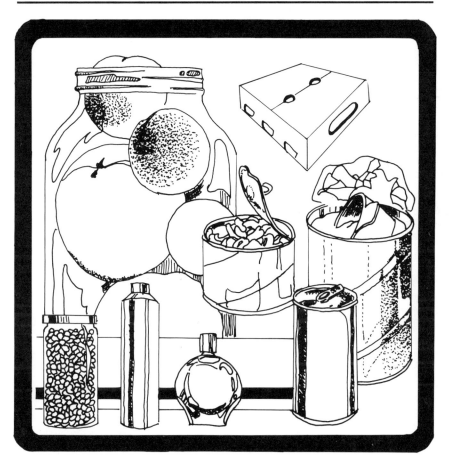

DONALD L. ABBOTT

Michigan State University

 KENDALL/HUNT PUBLISHING COMPANY
2460 Kerper Boulevard   P.O. Box 539   Dubuque, Iowa 52004-0539

Contents

Introduction

This text was written for use in an introductory course in packaging for high school seniors and college students. It will also be of use to industrial employees working in packaging that don't have the benefit of professional packaging training.

The text was written from notes used by the author for a number of years to teach a 3 credit quarter term course entitled, Principles of Packaging, at Michigan State University. Charts, graphs and illustrations have purposely been left out to avoid rapid loss of currency and to allow the lecturer to provide up-to-date data and statistics. Also many areas were left undeveloped to allow lecturers to expand on these areas as they see fit.

This text would not have been possible without the generous contributions and advice of my colleagues at the School of Packaging at Michigan State University, to which I am grateful.

CHAPTER 1

The Vocation
of Packaging

Packaging has been defined by the Packaging Institute USA, in their GLOSSARY OF PACKAGING TERMS, as the enclosure of products, items or packages in a wrapped pouch, bag, box, cup, tray, can, tube, bottle or other container form to perform one or more of the following major functions; (1) Containment for handling, transportation, and use. This was probably the original function. The basic reason for the package is to enable product movement. Packaging allows one to carry not only what can be held in a hand, but also products such as liquids, and dry free flowing powders, which are simply not transportable in consumer size units if they are not contained. (2) Protection and/or preservation of the contents for required shelf life, use life, and sometimes protection of the external environment from any hazard or contact with the contents. For example, most dry products are susceptible to moisture damage, and most wet products are susceptible to loss of moisture. Oxygen in the air can also affect certain products and the package can protect the product from it. For example, as a barrier, packaging retards the rapid oxidation of fat. Carbonated liquids such as beer, champagne and soft drinks use packaging to assure that the dissolved carbon dioxide is retained and the pressure maintained. Further, the combination of product in the inner and outer packaging must also provide protection against crushing from stacking and palletizing, and from breakage due to vibration and dropping, either in shipment or while waiting for the consumer to purchase and use it. (3) Communications or identification of content, quality, quantity, and manufacturer, usually by means of printing, decoration, labelling, package shape or transparency. The design and decoration must facilitate selection and motivate sales and, increasingly, they must meet government requirements including in-

1

gredients listing and nutritive value. (4) Utility or performance which would facilitate dispensing and use of products, including ease of opening, reclosure (if required), portioning, application, unit of use, multipacks, safety, second use or reuse and working features such as are found in aerosol sprays, cook-in-the-bag, memory-packs and especially provision for instructions or directions.

If the device or container performs one or more of these functions it is considered a package.

Packaging is also the development and production of packages (filling, closing, labelling), by trained professionals or operators employing methods and equipment designed for specific product lines in types of packages. Some 30 different categories of packaging machines are employed. For example, fillers, counters, cappers, labelers, wrapping equipment, cartoners, case loaders and sealers, as well as a wide range of support equipment for package making, inspection, monitoring, and handling. Increasingly, packaging machinery is designed and integrated to provide a complete system.

Packaging is often designated as consumer type, industrial type and military type. Consumer packages are intended for retail merchandising and emphasize sale features. They generally consist of large quantities of small units highly decorated. Industrial packages are intended for use in manufacturing, service and repair outlets. They are generally characterized as being large units or heavier units with little or no decoration. Communication on the package tends to be instructive only. Military packaging is highly specialized, very protective and requires extreme detail. The problem with military packaging stems from the fact that the users, the military branches of service, do not know when that item is going to be used, where it is going to be stored or under what conditions. Thus, packaging for military goods has to be the best that is available so that those items will stay usable for long periods of time.

History tells us that, originally, there was no requirement for packaging, that people were nomadic and tended to wander in search of food from one source to another. It wasn't until they eventually found more permanent type shelters that food then had to be gathered and brought to a common place of living. This, of course, brought about the first requirement of packaging which consisted of primarily simple things such as gourds, shells, leaves, and any other natural substance that was available and would contain quantities of foods or other items that were necessary to take to their homes. Anything, such as pieces of bark, hollowed rocks, hollowed tree trunks, and certainly

2

pieces of wood that were hollowed out were available and used. Some containers were made from animal parts such as bladders which were considered to be a good packaging systems because of their flexibility and capability of expanding. Animal skins were used for several years, and, in fact, are still used in some parts of the world for containment of beverages such as wine in Spain. It is known that Christopher Columbus used goat skins to contain water as he sailed towards the New World. Horns were a common commodity for packaging, some tend to be hollow and could be cut off from certain animals such as cows. Horns were used in the early days of the Revolutionary War for containing gun powder. Years before that, bones taken from large animals could be broken open, the center portion was cleaned out and, then used to hold certain items. In the tropical areas, bamboo was the common package because it grows in cells and is hollow. It is quite easy to separate these cells, make a small opening, and use them for a variety of different packages. Other parts of animals, such as sinew which is connective tissue was commonly used to bind and strap other packages and wrappings. Hair was known to be used as a woven material for making mats and cloth-like substances. Some of the grasses throughout the world have been and still are used to make mats and baskets. One author cites in his writings that, during the time span of 10,000 BC up to 800 AD, along with the natural derived materials such as cords, leaves and shells used by ancient man, were cups and containers made of human skulls. One of the things that archaeologists turn up most frequently in the domestic debris of Peking man's caves are human skulls, in which the foramen magnum, the opening through which the spinal cord connects with the brain, has been artificially enlarged after death to enhance use as a container.

Stone Age men had to make use of whatever nature provided when they wanted drinking cups or cooking vessels. In some areas there were gourds which are hollow vegetables. Near the sea large scallop shells could be used. Wherever man was to be found, there were skulls whose upper cranial section made an excellent and almost indestructible container. The practice of using skulls as containers carried on to the nomadic Scythians, who made containers from the skulls of people they particularly disliked.

About 6000 BC pottery evolved. Man had finally discovered how to make a continuous supply of fireproof and waterproof containers. Use of pottery is considered to be one of the most important achievements of the Neolithic period (6202–1400 BC).

Perhaps the widest application of pottery as a packaging material came about because of its usefulness in shipping and trading. The entire civilized world at the time depended on trading to maintain its food supply. Led by the Phoenicians, the Egyptians, Greeks, and Romans all traded wheat, wine, olive oil, spices, and perfumes. A container was needed to hold these items, and pottery appeared to fit the bill. Decorated and painted jugs and vases were used to identify contents and specific use.

During the ancient Grecian period or possibly earlier, pottery vessels with handles fitted to both shoulders and having pointed bottoms were pressed into the ground to store liquor. These vessels were also used for transporting liquor by attaching them to the saddle of a horse or to the frame of a ship. For dispensing the liquid, skin bags, earthenware vessels with pointed bottoms, or small vessels with rounder flat bottoms were used. Neolithic man (last period of the Stone Age) is known to be the first user of metal and pottery. Metals used were soft ones that could be found in near pure form, mostly lead, gold and silver. They could be pounded into foils and used not unlike the way we use aluminum foil today. The crude pottery that they used was simply made by shaping clay and baking it in an open fire. About 3000 BC to 4000 BC the Egyptians started making and using glass bottles and jars. These glass bottles were extremely small and had great value. They were used strictly for oils and scents that were also of value and these in turn were used only by rich people. The Egyptians first started to make paper from papyrus reed. They took the leaves of the reed, formed a mat on a flat stone, wet it down, pressed and dried it with another flat stone, laid alternate leaves on top of this (alternate being perpendicular) and continued to press down and smooth as they added moisture. Eventually, when they got the layers built up to sufficient thickness, they allowed it to dry in the sun. When dried, they could peel it off the flat surface and they did, in fact, have a crude form of paper.

During the late 18th century there was a rapid major change in economy with the introduction of power driven machinery along with advancement in technology. This was called the Industrial Revolution and it advanced container invention and fabrication significantly. Out of this Industrial Revolution came five important advances in packaging—(1) the metal can which is now the World's most used package of processed foods, (2) the collapsible tube which was invented by an American portrait painter for packaging portrait paints, (3) the folding carton which is said to have the most impact on self-service selling as we know it today, (4) the corrugated shipping case which has virtually replaced the wooden container as a shipping box, and (5) the crown closure which made an inexpensive and very efficient method of closing narrow neck bottles that contained carbonated beverages such as champagne, beer and soda.

During the late 19th and early 20th centuries, we brought mechanized production and food processing to packaging. We developed graphic arts for better communication on the package and started flexible packaging to reduce packaging costs and still provide adequate shelf life. In 1899 the Uneeda Biscuit Company started packaging their crackers in a unit size package which was made available to the consumer. This is considered to be the end of the cracker-barrel era. For the first time the consumer could get a reasonable amount of crackers in an individual package instead of going to a large barrel and measuring out a certain quantity. This brought on added protection of flavor and texture and started the use of individual packaging as we know it today. We now know that there has been more progress in packaging, storage, and distribution of goods in the last 100 years than in all of the previous times since people walked the earth.

It is interesting to note that, to certain peoples in the packaging industry, packaging is thought of in different ways. For instance, the consumer looks at the package as a product that is barely noticed, except when it fails to do it's job. It is generally purchased simply to be discarded. The manufacturer looks at packaging as a matter of cost and plant efficiency. The manufacturer looks at cost critically because about one-third to one-half of its operating costs goes towards the purchase of packaging materials and systems. The rest of cost goes to making the products and the general overhead of operating plants. The advertisers look at packaging as an expensive powerful promotional vehicle. It is known that the package itself will help sell the product. In fact, some authors call the package the silent salesman. Studies have shown that the package is probably the cheapest form of advertising because it will get more recognition with the least amount of cost when comparing it to other advertising media such as magazines, television, radio, etc. To distributors of products, packaging is something more to handle, stack and display. Hopefully it will assist the distributors in controlling and stacking these goods.

Mr. R. Bruce Holmgren, Editorial Director, PACKAGING MAGAZINE, outlines five requirements to make a perfect package. Mr. Holmgren states that first, the package must be attractive so that it wins the sale. If nobody buys it, we might as well stop and talk about something else. Second, it must perform properly and do its job structurally and functionally so that people will buy it again—otherwise the first sale is all for not. Third, it must be economical to produce so that the company turning it out can keep that package and its product in the marketplace. Fourth, it must be durable to withstand the ravages of distribution, now tougher and tougher with longer shipments as a result of companies using fewer shipping points. Fifth, it must meet a host of legal and regulatory requirements, as well as responding to a range of current pressures, such as consumer demands and the problems of disposability. His final comment is, if the package fails on any one of these five points, it is not a perfect package.

Today packaging involves nearly all of 200 industries, 300,000 manufacturing companies and countless service fields in the United States. It is estimated that about one million people are employed in packaging, making it the largest single employer of people in the United States. We know that about 75% of all goods produced in this country are packaged at some point in their distribution system. In recent years packaging has consumed some 55 billion dollars of the value of all finished goods sold in the United States, and it is estimated that by 2000 total packaging sales volume will approach 70 billion dollars. The packaging industry today uses 50% of all paper and paperboard manufactured, 45% of all glass manufactured, 37% of all inks manufactured, 29% of all plastics manufactured, 25% of all adhesives made, 23% of all the aluminum made and 8% of all the steel made in this country. Packaging has been described also as the country's largest single industrial user of paper and glass and the third largest user of steel, the third largest industry and in the U.S.A, and the largest employer of people in the United States.

The components of packaging are of four different categories. First is forms. The form of packaging is broken down into three types. Number One is the primary package. The primary package is one that is in direct contact with the contained product. It provides the initial and usually the primary barrier protection. Examples of primary packaging includes cans, glass and plastic bottles, pouches, etc. The primary package is the complete form that actually contains the product. Number two is the secondary package. It is always on the outside of the primary package. In much of the industry, the secondary package is a consumer package that unitizes two or more primary packages into one. For example a six pack carrier of beer or soft drinks is a secondary package and a paperboard tray plus film overwrap for 10-unit portion packages of dry breakfast cereal is a secondary package. Many products do not have a secondary package. Number three is the tertiary package. It is the physical distribution carrier. Usually this is the corrugated case containing a number of primary or secondary packages. In the last few years, although corrugated or paperboard has remained the dominant form, corrugated trays plus plastic shrink film, shrink film alone, kraft paper bundling, etc., are being increasingly used for this application. The tertiary package offers some protection although, in some instances, protection is more in the nature of tying together the primary or secondary packages so that the strength derived solely from their union is maximized. In general, the tertiary package is the carrier not seen by the general public.

Second is materials. Packages are composed of materials whose function is to contain or separate the product from its environment. The major materials used in packaging today are paper and paperboard, glass, metals, plastics, and wood. There are many other supporting materials such as adhesives and inks.

Third is machinery. Machines used in packaging could include those machines that are used to convert raw materials into packaging materials. However, two distinct classes of machinery are used in packaging. First, we use the machinery for converting materials into form containers which are transported to the packager; and second, machinery for subsequent filling and closing the package.

The fourth component of packaging is people, the single most important component. Despite the vast array of automatic and semi-automatic equipment, a large fraction of packaging is still performed manually or, at best, with mechanical assist. All, however, must be serviced, supervised, monitored, or otherwise managed by people.

Today the annual per capita consumption of packaging materials is estimated at 594 pounds. This has remained pretty much the same for the last several years and is expected to be the same in the future. To break that down into individual containers, each one of us consumes over 1,000 paperboard boxes, 419 metal cans, 338 closures, 262 beer cans, 137 soft drink cans, 185 grocery and merchandise bags, 170 glass bottles, 115 cigarette packages, 76 plastic bottles, 63 milk cartons, 32 fiber-foil cans, 20 liquor and wine bottles, 10 aerosol cans and 8 collapsible or squeeze tubes for a variety of products.

CHAPTER 2

Materials of Packaging

This chapter is broken down into sections—one section for each of the four major materials used in packaging. The major packaging materials are glass, paper, metals and plastics. Wood has been included although most packagers don't consider it to be a major packaging material but more of an aid in package distribution. But since so much is used in the package industry, it is included in this chapter.

Section A: *Packaging in Glass* (This section, courtesy Brockway Glass School Manual, 1986)

Glass is a unique material. It has the molecular structure of a liquid and the physical characteristics of a solid. It has sometimes been referred to as a super cooled liquid. A more complete definition is that glass is an inorganic, noncrystalline solid, formed by cooling from the liquid state which shows no discontinuous change at any temperature, but which becomes more or less rigid through a progressive increase in its viscosity. Glass is one of the oldest substances known. A material that is very similar to glass can be found in areas of volcanic action. This glass is sometimes thought to be used as arrowheads during the Bronze Age. Some historians have dated glass beads back to 12000 BC. We also know that at about 7000 BC Egyptians were using glass amulets as symbols to protect the wearer against evil. The first man made glass was made unintentionally when some sailors went ashore at a deserted sandy island. They used some blocks of soda from their cargo to make a fireplace on the sandy beach. Sometime later, after the fire had gone out and cooled, they were disassembling the fireplace to return to the ship. They discovered that some of the soda had fused with the sand and formed glass. Around 3000 BC historians tell us that the Egyptians were using glass sparingly. Some of these glass objects can be found today in the London Museum. The containers made at that time were mostly formed from molten strands of glass that were spirally wound to form containers. They were very fragile, very small, very expensive, and used only by the very wealthy to contain perfumes and oils. By 1550 BC glass had become an important industry in Egypt. It was still very expensive and only available to those who could afford it. In 300 BC the first glass blow pipe was used. This was considered to be a big breakthrough in glass container manufacture. For the first time people could make large sized containers with a fair amount of quality and precision. The task of making these containers was very involved usually employing up to seven people consisting of three craftsmen and four boys that would assist these craftsmen. The last portion of the container to be completed was always the opening. Because of that, the term "finish" became associated with this opening. Even today the term "finish" denotes, on most glass containers, the opening or threaded part of a glass container. By the third century, we know that Romans were starting to cast window glass on flat rocks and be able to use some of this glass in construction. The glass remained extremely expensive until about the 18th to 19th century when machines came on the scene and were able to convert molten glass into usable containers at a fairly fast rate. This brought prices down and made glass containers available to the average person. In 1903 the first fully automatic bottle making machine was used. It was made by M.J. Owens in a Toledo glass plant owned by E.D. Libby. The name "Libby Owens" in glass manufacturing is well known today. The effect of the machine manufactured glass container was incalculatable. The evidence was everywhere. Cheap bottles modernized milk distribution. Heretofore, sanitary milk delivery conceived in the 1880's by a Dr. Thatcher in Potsdam, New York had been hampered by high costs. Typically in this time, the milkman came around with a dipper for delivering milk from a large can. Dr. Thatcher's prescription for getting no dirt in the milk was to adopt a wide mouth, easy-to-clean bottle and a closure with his nickel plated lightening type clamp that pivoted on the neck. But not until 1910 did the quart bottle of milk become an American way of life. With machine made bottles, a complete package was no longer expensive. Today more than 40 billion glass containers are produced in this country each year. There have been problems in recent years such as labor problems, price increases, and energy problems, that have hurt the glass industry by accelerating the change from glass to plastic in many packaging systems. Today food packaging uses well over 12 billion glass containers. Beer packaging uses about 11 billion containers, soft drinks about 9 billion containers, wine and liquors about 3.5 billion containers, health care products about 2.5 billion containers, cosmetics and toiletries about 1.3 billion containers and household items about one-half billion containers. The future growth of glass industry will depend primarily on the competition from plastic containers. As the raw materials of polymers fluctuates with the supply of petroleum, glass use will be influenced also. Some people believe that there is a slight trend back to glass, however it is felt that the impact will be insignificant.

When we look at the composition of glass or common glasses such as those used to make containers, they are composed of a combination of various oxides. Some of these oxides will form a glass on their own and are known as network or glass formers. The most common of these is silica. While silica alone can be used to form a glass, it has some features which make it undesirable for container glass. Some of the advantages and disadvantages of the pure silica glass are as follows. Advantages are—inexpensive, single component, low expansion, and good durability. The disadvantages are—it's very hard to melt and it's very hard to form, thus making it very expensive. In order to overcome some of the disadvantages of a purer silica glass, other oxides are added. The most common of this is soda. As with pure silica glass, a soda silica glass has advantages and disadvantages as follows. The advantages are—it gives a lower melting temperature and a lower viscosity. The disadvantages are—soda is expensive so the cost goes up. You do get a higher expansion rate and you do decrease durability, but the durability is the chief disadvantage of the silica-soda combination. With the right combination this type of glass is actually soluble in water and is commonly called water glass. In order to overcome the poor chemical durability, another class of oxide is added to put some strength back into the network. The most common network stabilizer is calcium. The results of adding calcium oxide are as follows. The advantages are, a lower expansion rate and good durability. The disadvantages are, an increased melting temperature and an increased viscosity. The addition of the calcium dramatically improves the chemical durability and it also enhances devitrification which is the tendency of a glass to revert to a crystalline structure. Another oxide is added to overcome the devitrification tendency. The most common of intermediate oxide is alumina. The

11

results of an addition of aluminum trioxide are as follows. The overall advantage is lower devitrification and better durability. The disadvantage is higher viscosity. Some other modifiers can be added at this time but they are considered to be minor and will not be covered in this text. The actual composition of a typical glass container is silicone dioxide 70%, sodium oxide 15%, calcium oxide 12%, aluminum trioxide 2% and the minor additions accommodate 1%. The properties from this combination for glass containers produces a moderate cost container, low viscosity during forming, chemical durability, inherently strong, low thermal expansion, nonpermeable, tasteless and odorless, a clear transparent glass, or if we want UV light protection, we can select an amber color glass. Almost all glass made in this country has cullet added to it. Cullet is simply defective glass or recycled glass that is ground up and added to the mixture of sand, sodium, calcium, and aluminum. It has two purposes or two benefits, which are to speed up the melting action and to recycle or reuse glass, both of which will reduce costs. In this country about 35% of all glass manufactured is made up of recycled glass or cullet.

Some of the amber glasses, such as in beer bottles, the percentage of cullet used is about 70%. When we look at the overall raw materials which supply the major components, and summarize what their advantages are, we can see that sand is an abundant material with a moderate cost, usually plentiful, near the glass manufacturing plants. The soda ash is expensive. There are only two deposits in the United States, one in Wyoming and one in California. The limestone is an abundant material at a moderate cost and it can usually be found near the glass manufacturing plant. The aluminum trioxide, or feldspar, is primarily deposited in North Carolina, and as only a moderate or a very small amount is used, it adds only a moderate increase in cost. The manufacturing of glass requires a large factory with a large amount of capital to build the various furnaces, mixers and forming machines. The raw materials usually are combined in what's known as a batch house above the furnace. Once they are mixed, usually some water is added to decrease the possibility of segregation and enhance the pattern flow into the furnace for efficient melting. The batch, once checked for quality and quantity, is charged into the furnace where it's melted into glass. The batch of raw materials is charged into the melting furnace at the same rate as glass is being pulled out at the opposite end. So the amount of glass in the furnace is kept constant at all times. The furnace consists of three principle parts: the melter, the refiner and the regenerator or checkers.

The melter is a rectangular basin in which the actual melting takes place. Along each side of the melter above glass level are from three to six ports. These contain natural gas burners and direct combustion of air and exhaust gases. The melter portion of the furnace is separated from the refiner by a bridge wall. This is a shallow wall which has openings on the bottom (throat) and separates the melter atmosphere from the refiner atmosphere. Glass passes from the melter to the refiner through the throat which is a water cooled tunnel through the bridge wall. The refiner acts as a holding basin where the glass is allowed to cool to forming temperatures before entering the glass formers. So the bridge wall is simply a wall which separates the glass in the melter portion of the furnace from the glass in the refiner portion of the furnace. At the same time, since glass has to flow through a throat, all the impurities in the mixture will be held back because they float to the top in the melting portion of the furnace. These impurities then are periodically skimmed off.

The temperature ranges for the various glass forming functions are: (1) 2700 degrees fahrenheit to 2400 degrees fahrenheit for melting. (2) In the refining portion of the furnace, the temperature runs from 2400 to 2100 degrees fahrenheit, and (3) during the forming phase of glass container manufacture, temperatures will range from about 2000 degrees down to about 1400 degrees fahrenheit. As the glass exits the furnace going into the glass forming machines, it is discharged in small shapes called gobs. The gob is simply a measured quantity of molten glass which will be formed into a glass container. The size of the gob will vary with the size of the container. Typically a gob will vary from one-half ounce to forty-eight ounces in weight. Seven ounces is a typical weight for a twelve ounce beer bottle. Temperatures of the gobs are usually maintained at about 2100 degrees fahrenheit and has the consistency of thick honey. Speed or discharge of gobs from a single neck port can vary from 30 gobs per minute to 150 gobs per minute depending upon container size produced and forming machines used. Interesting enough, the shape of a gob is also controlled because there is an optimum gob shape for each glass container produced. The shape will effect the precision of the gob entry to the forming machine. These shapes are controlled by plunger, height, stroke, size, shape, machine speed, temperature, orifice size, sheer cam, and tube height. Once a gob leaves the furnace, it drops by gravity into a scoop which routes the gob to a section of the glass forming machine that is ready to receive it.

This scoop moves constantly, and constantly feeds the gobs as they leave to different molds. Usually, by the time the gob is delivered to the glass mold, it is about 2000 degrees fahrenheit. Container forming machines can possess anywhere from four to ten molds. Production rate of a glass molding machine will vary from 30 to 530 containers per minute. Total time required to produce a container varies. For the typical beer and soda pop container it takes only about 10 seconds. The first form of glass as it goes into the forming machine is called a parison. A parison is defined as a specifically shaped formation of glass which will be blown up like a balloon, or more specifically, the parison is the premature shape of the glass container. It does have the finish, the opening and the threads in its final form. It is hollow inside. It's temperature usually ranges down to around 1700 degrees at its skin and, obviously, it has the same amount of glass as the container which it will produce. The purpose of forming a parison is to maintain quality distribution of the glass for the final container. Typically created in just two seconds, and since the glass is still at a high temperature it is just barely able to hold its shape. This shape is immediately transferred to the final mold where high compressed air is blown in through the finish causing the parison to expand again like a balloon and conform to the shape of a final mold. Within seconds this mold separates and the container is firmly but carefully extracted by the use of transfer tongs that place the final shaped container onto a conveyor system that has cooling pads to start the cooling process. When properly cooled, the glass containers will get a surface treatment. The reason for this surface treatment is that glass containers will be subject to contact with other containers and conveyor equipment during travels throughout the remainder of the manufacturing, shipping, and filling operations.

The strength of the outer surface of the container is maintained with the application of surface coatings. The principle function of surface coatings on glass is to provide lubricant to the surface. The lubricant permits the container to rub against line rails or other containers without causing the glass to be degraded or weakened. For years glass manufacturers and users have known that glass has a tremendous strength potential. From experiments involving glass fibers drawn under ideal conditions, the maximum strength of glass has been estimated to be as high as 1,000 to 3,000 kilograms per meter squared. However, manufacturing defects and subsequent abrasions during handling reduce the usable strength of commercial glassware to the range of 10 to 100 kilograms per meter squared.

Some glass experts claim that pound for pound new glass is equal in strength to steel. It is well known that glass always fails from tension on the surface providing it is otherwise free of internal abrasions or faults. Severe surface abrasions can dramatically reduce the strength of a piece of glassware. As a result many approaches have been proposed over the years in an effort to form a protective surface capable of protecting glass from normal abrasion. By such means the strength of the normal glass container can be maintained closer to the upper strength potential.

The first coating used by the glass container industry was simply sodium sulphate. It was formed on the outside of the glass articles while they passed through an annealing lehr into which sulphur dioxide gas was being bled. A coating would act as a mild lubricant until it was washed off the containers. This process was discontinued because of its bad effects on the annealing lehr and its environmental problems. Types of surface treatment materials available today fall into two basic categories. The first category is a nonpermanent coating such as PVA, polyethylene, green soap and stearate. These types of surface treatments are normally used on glass containers for cosmetics, shelf stock, small capacity food items, returnable beverage bottles and many other containers. The second category of coatings is permanent coatings used to form a tough abrasive resistant coating. For the most part these coatings are permanent in that they are not affected by wet processing as much as the nonpermanent coatings. They include silicone and the tin or titanium dual coatings such as tin/polyethylene, tin/PVA and tin/stearate. The purpose of any of these surface treatments is to prevent surface abrasions which produce damage and weaken the container. They all require two properties which are lubriciously and abrasive resistance or the ability of the coating to reduce the coefficient of friction of glass surfaces in contact with other glass containers or with handling equipment and the mobility and protection against damage under low contact pressure.

Once formed all glass must be annealed. The purpose of annealing is simply to reduce or relieve structural stresses that occur during rapid cooling after the glass exits the furnace. Annealing is simply a controlled heating and controlled cooling process. It is designed to relieve internal stresses introduced in a glass container during and immediately after glass container formation when the glass undergoes a tremendous change in temperature. The process is accomplished by conveying these containers continuously through a large oven called a lehr. Upon entry into the oven, heat is applied bringing the glass containers up to about 1050 degrees fahrenheit and holds this temperature for about 15 minutes. This time allows the glass container to relax and remove all of the stresses that occurred during forming. After this 15 minute holding period, the containers are then slowly cooled back down to near room temperature by the time they exit the lehr. Times for accomplishing this operation vary significantly depending upon the thickness of the glass walls. It can take as little as 20 minutes or as much as 90 minutes to complete the annealing process. Without an annealing process, glass containers would be completely unsafe because they would easily break, and would be unsuitable for packaging operations requiring conveying systems and handling. It is basically impossible to obtain unannealed glass for study from the glass manufacturers because they are responsible for any injuries that occur from the shattering glass. Once annealed, the glass containers are ready for shipment. If it is necessary, they can be labeled or printed upon at this time. The printing processes are varied. They can be silk screening, enameling, etching, sand blasting or coating. There are many decoration opportunities for glass containers. Glass containers are generally shipped on pallets. Machines load the pallets by tiers. Once the pallet is loaded, it is secured and sent off to a holding area for distribution to the customer.

The key bottle parts or definitions of bottle parts are: (1) the finish which is the part of the bottle for holding the cap for closure, sometimes called the opening of the container. (2) the neck which is the part of the bottle between the finish and the shoulder. (3) the sealing surface which is the surface of the finish on which the closure forms a seal on the very top of the container. (4) the shoulder which is the portion of the glass container in which the maximum cross section, or body area, decreases to join the neck of the container. (5) threads, which is a small spiral protruding glass ridge on the finish of the container intended to be sealed with a screw type closure. (6) a transfer bead which is below the threads and is a protruding ring of glass on the lower part of the finish used by machines to transfer the glass container from the mold to the cooling pad on the blowing machine. (7) the body which is below the shoulder and is generally straight sided, but can be any shape the user wants. (8) the pushup is at the very bottom. The purpose of the push up is to add internal compression strength to the glass container and also give it stability when sitting on the shelf or conveyor. It forces the glass container to rest on a circular ring-like portion on the bottom of the container called the bearing surface. (9) the whole bottom portion of the glass container is called the heel.

Glass containers can be ordered in most colors—red, yellow, green, blue, violet etc. Colored glass requires a chemical coloring agent to be added to the furnace batch. Certain colored glass will protect the contents of a glass container from light in varying degrees depending upon the color. The light that causes damage in consumable products is ultraviolet light. There are three colors that will do a fair job of screening out ultraviolet light. The most common color and the most effective is what is called amber or sometimes referred to as brown bottles. This color can be obtained by introducing a carbon sulphur mix into the batch and is commonly associated with beer bottles and pharmaceutical bottles. Another color that will do a fair job of screening out ultraviolet light is red. Red can be obtained by adding cupric oxide or cadmium to the batch mix. It doesn't do as good a job as the amber and tends to be more expensive. Therefore it's not too popular from an economical point of view. Another color that does some screening is green. This is obtained by adding ferrous sulphate to the batch. It gives a nice color, but the effectiveness is diminished and has not been very popular with most packagers. All other colors are not known to do a sufficient job of screening and would be used strictly as a marketing device by the packager.

There are basically seven types of glass containers used in packaging. (1) Bottles are the ones that are most used. They can be any shape you want but are specifically identified as having small openings and narrow necks. Sometimes the necks are long as in a wine bottle and sometimes they are short and stubby. They are mostly round and are generally restricted from about four ounces to one half gallon size. They are used mostly for liquids because of the small opening. They will accept many closures, from corks to press-on crown caps to screw on caps. The bottle is generally known as the pressure package because it excels in packaging carbonated beverages. Typically a one quart size bottle would cost the packager somewhere between 10 and 15 cents each. Some 31 billion bottles are manufactured and used each year. (2) The second container is a jar. It is identified by the fact that it has a wide mouth designed to facilitate filling, especially of solid foods such as whole peaches. It also facilitates getting the product out by inserting instruments into the container assisting the removal. Commonly used for food and drugs, they obviously have large closures. In recent years these closures have been pretty effective and are not the limiting factor that they

17

once were. The big advantage is easy filling and convenient dispensing. Typically, a one quart container would cost somewhere between 7 and 10 cents apiece to the packager. About 9 billion jars are made each year for the packagers. The rest of the containers we mention are not used in near the quantity as the first two so total numbers will not become a factor. (3) The number three container which can be used in packaging is called a tumbler. A tumbler is simply a water glass or a drinking glass beefed up a little bit and commonly used as a marketing device to stimulate sales for jams, jellies, preservatives and other types of spreads. They are characterized by the fact that they have no neck. They can be straight sided or taper up to a larger opening than the bottom. They use a clenched friction overcap commonly assisted in sealing with a vacuum. Sometimes you can get a plastic reclosure to go along with this because, once the container is opened, it is very difficult to reseal it again because of the damage done to the metal friction overcap. They are usually pint size or less. The big advantage is the value in reuse and for special promotions because they can be decorated to attract the consumer's eye. (4) The fourth container used in packaging is called jugs. These are usually gallon size but can be as small as a half gallon and as large as to two gallons. They are characterized by the fact that they are large heavy glass containers, usually with handles commonly known in the past as cider jugs. They are used extensively for liquid chemicals and for institutional foods such as vinegar and ketchup. Typically a one gallon size would cost the packager in the neighborhood of fifty cents each. (5) Some larger containers are called carboys. They are made of very thick glass making them very heavy in weight because they are expected to have a very long life and many reuses. Common sizes are from 3 to 13 gallons, but the most popular size is in the 5 gallon capacity. They are commonly used for drinking water in offices and factories and also for chemicals in chemical plants, universities, and institutions of teaching and research where large quantities of chemicals are used. They are not commonly found in the home because of their size and difficulty in carrying and handling. (6) Our sixth container is what we call a vial. It's a very small container. It has a flat bottom so it sits nicely on conveyors or in storage. It usually has a rubber stopper in the finish area and is designed for various drugs. This stopper allows medical personnel to insert a needle syringe for the extraction of either drugs or chemicals that are used in small amounts. The advantage is that the container is self sealing and multiple extractions are possible. (7) The last container is an ampule.

These are very small, very thin walled and usually a maximum of one-half inch in diameter. They are made from glass tubes. They have a very narrow neck and a hour glass shape. Once the container is filled with a sensitive chemical or pharmaceutical product the top portion of the glass is sealed by flame. It's a complete glass seal all the way around used for portion packaging because, to get to the contents, one must destroy the package or actually break the container in half at the narrow neck thus allowing extraction of the product. Some uses of ampules are for expensive essences, food ingredients, cosmetics, perfumes and for very sensitive drugs. Most drugs are single dosage injectables.

The packaging industry likes glass in packaging because, from a filling standpoint, glass containers handle very well by machines. They also store, stack and fill very well. In fact, filling speeds for glass containers are easily done at speeds from 1,200 to 2,000 containers per minute. They can also be filled slowly so for a small operation they still handle very nicely. Cost of glass at this time is favorable to other containers made from metals and plastics. Glass has a property of inertness, it does not react with most food and drug items. It gives excellent visibility to the product which promotes impulse selling. It has high strength, excellent rigidity, and provides one of the better shelf lives for product protection. The glass doesn't deteriorate or dent and reclosure systems on glass containers are very good for most sizes and shapes. Glass does enjoy a very good recyclability rate. Also, glass containers can be reused over and over again sometimes as many as 15 to 20 refillings of products. Glass is microwavable which is a popular property in today's market. It can be creatively shaped or decorated to meet the customers' needs. It makes an attractive package that is accepted by the consumer and can be formed in a variety of sizes and shapes with little limit to the styling of glass containers.

Glass has some disadvantages. Otherwise there wouldn't be any other containers on the market that could compete with it. The disadvantages are two. First is weight. Glass containers are very heavy, and although there are many factories throughout the world and in this country, containers almost always have to be shipped a long distance to get to the packager. This requires the shipping of a lot of weight and a lot of air or volume to get the container to the packager as opposed to plastic containers which can be blowmolded in the factory just prior to use and/or paper products and cartons which can be assembled or structured just prior to filling. The second drawback to

19

glass is fragility. Glass containers must be handled with a certain amount of care to reduce breakage. There's always the risk that if there is a slight defect in a bottle of carbonated beverage it could literally explode. This has happened many times in the past especially with a larger 2 liter size bottle.

From a manufacturing point of view there is a tremendous capital investment to equip a glass manufacturing factory, and there is a tremendous amount of energy required to form the glass containers from raw ingredients up to annealing. Still, the future use of glass containers in packaging looks good. Glass sales go in cycles. At the current time there tends to be a little bit of a rise in use of glass containers due to marketing changes, energy changes and raw material changes. It is expected that use of glass in packaging will enjoy a very slight growth over the next ten years. Not as much as some of the other materials but still a positive trend. As manufacturers develop newer techniques to cut costs and provide certain advantages to glass containers, they may enhance the use of glass containers throughout the packaging industry.

Section B: *Metals in Packaging*

The use of metals in packaging dates back to almost as far as man goes. We know for instance that metals that could be found in their near pure form such as gold and other metals that were malleable could be beaten into shapes and used as cups and bowls. Even wrapping materials such as lead foil could be made. Some of these materials were still being used as a wrap as late as 1926 when lead foil was and had been in use to line tea boxes for ocean shipping. As late as 1930 lead foil was used in cigarette packaging. The use of metal cans which is now a primary packaging form was encouraged by Napoleon in about 1809. He was having problems with feeding his army which was forced to consume spoiled and inadequate rations during their trips. His Navy developed scurvy from the lack of proper foods. He offered the sum of 12,000 francs, which was a tremendous amount of money at that time, to anyone who could develop a means of packaging or keeping food consumable over long periods of time and that could be shipped many miles. In 1809 a Frenchman named Nicholas Appert found that food which was sufficiently heated in a sealed container would not spoil. His primary method or vehicle for packaging was a glass container. This satisfied the requirements of the reward money which he did collect but the glass container was not suitable for armies and further work had to be done.

In 1810 an Englishman by the name of Peter Durant came up with what was called a tin canister. It was sheet steel formed into a cylinder which then had tops and bottoms sealed to it by the use of solder. This was a more suitable container for Napoleon. But the method of manufacturing left a tremendous amount of lead solder on the inside of the can which later caused many health problems, but at the time was considered a breakthrough and was put into use. It wasn't until some 50 years later that Pasteur explained why the system worked and what these two gentlemen were really doing in the preservation of foods. At this time a typical tinsmith could make somewhere between five and six cans per hour. Metal cans were used by Admiral Byrd when he made his trip to the North Pole in 1824. More than 100 years later a return trip by other explorers found some of these cans intact with food that was still good. In the USA during the 1850's the Borden Company started to can condensed milk because they were looking for a method of reducing infant mortality from contaminated milk. This one act alone through the use of canned and pasteurized milk significantly lowered infant mortality and thus gave a very positive image to the metal can. Today, typically 27 percent of all the packaging done in this country is in metal containers. Packaging metals consume a total of 7 percent of the steel production and 21 percent of the total aluminum production in the United States. The containers are mostly cans and typically break down something like this. Beer sales account for about 37 percent of the cans made, soft drinks 32 percent, foods 27 percent and nonfood items such as oils, paints and aerosols 4 percent for a total of well over 100 billion cans per year used by the American Packaging Industry. At this writing there were 168 companies making cans and 397 plants throughout the country. At the same time, we use 35 million metal drums for chemicals, paints, inks and oils and some 71 million pails for things like adhesives, paints, inks and oils.

Steel is made from iron ore which is an abundant substance on the earth. It has been in use for use for over 4,000 years. It is found in most rocks and soils and it must be assimilated by people and plants. It is a constituent of hemoglobin. Steel is simply identified as a very low carbon iron which gives it the property of being more versatile than elemental iron and far more stronger. Aluminum on the other hand is a relatively new metal first put to use in about 1825 at a tremendous cost per pound. A pound of aluminum at that time cost about $550 by the 1800 dollar standards. It is known that Napoleon used aluminum for forks and spoons for his special guests. Other or less important guests had to use gold or silver forks and spoons. It wasn't until 1888 that a less expensive ore extracting process was developed. This was partially due to the availability of electricity and other machines. By 1942 the price was down to 14 cents per pound so it started to be competitive after WW II and today is the growthy metal in packaging. Cans are the largest single type of container made out of metals. Metal cans are the primary package for storing, transporting and merchandising a variety of goods and substances including the bulk of the developed world's processed foods and beverages. They come in a large variety of sizes and construction for all types of goods from dry products to mechanical parts or machines. In 1981 for the first time the use of aluminum cans exceeded the use of steel cans. In this country 60% of the rigid metal containers are cans. The biggest use is beverages, then foods, then nonfoods. Secondly barrels and pails are used for liquids, semi-solids and dry products. They are primarily for industrial or chemical use and not commonly found in the home except for smaller paint cans or pails. Thirdly, a big use of metals is for foils. The fourth use is for collapsible tubes which is mostly aluminum and/or tin. The fifth use is for closures such as bottle caps and lids. The sixth use is for laminates or specifically aluminum foil laminated to plastics and/or paper or both.

As mentioned earlier, originally can manufacturing was very slow by hand. Typically a worker working a long day could produce 300 cans per day as late as the late 19th Century. Now machines manufacture 300 to 600 cans per minute or 288,000 per an eight hour day. The common three piece can starts with a flat blank that is cut and notched, rolled and hooked and then the side seam, after being hooked, is soldered. More recently it is welded and in some cases simply cemented or glued. The bottom is flanged and a lid is rolled around the can bottom and then soldered to complete the seal. After filling, the top is put in place and is rolled on with a double seam. Usually there is a layer of gasket material that has been flowed into the tops so that soldering of the top is not necessary. There was little change in three piece can manufacturing until the 1960's when they started to get a convenience closure of aluminum that could be peeled off. Can manufacture went to thinner steel on the side walls and concentric rings in the side walls. They then went to tin free steel for coating which significantly reduced the cost of a can because tin is in short supply and is expensive. By eliminating tin and putting on other types of coatings such as chromatic films or aluminum enamels and vinyls, the overall cost of the can went from 13 cents down to 8 cents. At the same time the thickness in the side walls decreased from 8 mils to 6 mils. This made the can very competitive and allowed it to enjoy a large market. Todays trend is to the two piece can. The purpose is to eliminate the side seam which has always been a weakness in metal cans and also to eliminate the separate bottom piece. This method is used today for the manufacture of both cans and tubes. It is called a draw and redraw where a slug or a small disk of metal is placed into a female die and then forced or extruded up the sides of this die by a male die impacting with sufficient force to cause the metal to flow. This makes a tube-like structure which then is trimmed and coated for further use.

There has also been a trend in beverage containers to neck down containers. This applies to aluminum containers and the net result is that it allows the top portion or the easy opening can top to be much smaller than normal. The separate can top with its easy opening device is an expensive aluminum alloy. Reducing aluminum can top diameters allows significant cost reduction. In fact it's well known that the body of the can is less expensive than the top even though there is far less metal in it. The advantage that aluminum has over steel is that it's far lighter in weight. It is a softer material that allows the easy opening or pull tabs. It can be made thin enough so that it becomes a flexible material called aluminum foil. Aluminum foil is the very best flexible barrier material available, and has a tremendous advantage in appearance. It also is a nonreactive metal but is still expensive. It does require protection as it is easily scratched and/or torn. Aluminum foil must be buried (laminated) within the package to protect it from outside abrasions or damage from machinery and other packages. The fact that it is a perfect barrier of moisture and gases is limited to thicknesses of greater than one thousandth of an inch. Thickness less than one thousandth of an inch thick or one mil will produce pinholes and destroy the barrier proof advantage.

The advantages that the metal can enjoys are that it has excellent product protection. It lends itself to high speed filling and casing at speeds up to 2,000 containers per minute. It provides good display by retailers through labelling and/or printing on the can and it's easy for consumers to store and use even though some cans do require a mechanical opener, which are common instruments found throughout the society. It does have disadvantages which are that the empty can shipping and storage takes up a lot of space. Once in storage cans must be protected from moisture and/or humidity changes because metal cans will stain easily from moisture and once stained are not suitable for packaging. Cans on the shelf do not maximize shelf space. There is a lot of air space between each can because of their round shape. Steel cans do not enjoy a very good recycling record. Aluminum cans have a very good recycling record because the aluminum industry supports it extensively. In fact it depends on significant numbers of used aluminum cans to cut down their converting costs. Considerably less energy is used to remelt aluminum cans than is to extract aluminum from its ore. Therefore it is a definite advantage for the aluminum companies to recycle cans which is not true for steel.

Today there are about 2,500 different products being packaged in cans—everything from soup to parachutes for the military service. A recent popular use of metal in packaging is for metallized films. Metallized films are paper and/or plastics that have a metal coating placed upon their surface. This coating gives the plastic and/or paper a rich appearance and improves its barrier properties thus gaining two advantages at a less cost than laminating aluminum to these carriers. Metalizing is accomplished in a vacuum chamber where the carrier film is slowly unrolled and rerolled within the vacuum chamber. Aluminum wire is subjected to high heat which causes a vaporization of aluminum without burning. This vaporized aluminum is directed to the surface of the plastic or paper as it is unrolled and rerolled within the chamber thus allowing a very thin layer of aluminum to be applied to the walls. Metalized film is used in flexible packaging for candies and snack foods.

Section C: *Paper in Packaging*

Paper is defined as the name of all kinds of matted or felted sheets of fiber (usually vegetable but sometimes mineral, animal or synthetic) formed on a fine wire screen from a water suspension. Paper derives its name from the term papyrus, a sheet made by pasting together thin layers of an Egyptian reed used in ancient times as writing material. Paper is one of the two broad subdivisions of paper, the other being paperboard. The distinction between paper and paperboard is that paper is lighter in basis weight, thinner and more flexible than paperboard. Paper's largest uses are for printing, writing, wrapping, and sanitary purposes, and it has a variety of other uses.

Paper as we know it today was first produced in China by Lee Yang. The early process of papermaking involved shredding the bark of mulberry trees, mixing it with scraps of linin and hemp then beating the mixture into separate fibers. Water was added to produce a pulpy mixture. He then dipped a mold of bamboo with a cloth floor into the pulp to form a sheet of paper on it. The paper was placed in the sun to allow it to dry. When dried it was removed and ready for use. The paper replaced bamboo strips and silk which was in use at that time for writing material. Paper and papermaking remained a secret in China for several hundred years but during the eighth century the secret was discovered by Moslems who had invaded and swept across China taking the papermaking process with them. The Moslems had captured a complete paper mill with laborers intact and took it all back to Baghdad in the year of 793. From there, the process eventually spread to Western Europe, Spain in 1151, Italy in 1390, France in 1348 and eventually to the United States in about 1690. England is reported to have started manufacturing paper in large quantities during the 16th Century and supplied all the needs of the colonies for many years. During most of this time there was really not much of a demand for paper because it was used strictly for writing. There were no printing presses and very few people were capable of writing or reading. It wasn't until about 1500 when the printing process was developed that the demand for paper started to rise to the point that there was a shortage of raw materials to make paper which in Europe was almost all rags. This shortage lasted until about 1800 when there was a shift from rags to wood fibers.

The use of wood fibers to make paper was discovered and reported by a Frenchman named R'eaumur who observed the paper wasp at work chewing up partially decayed wood fibers and making a nest out of it. The nests when examined were really several layers of paper, gray in color. R'eaumur felt that if the insects could do it, people ought to be able to do it and he started working on a method of using wood fibers. By 1830 there was a shift from rags to wood fibers. Wood at that time was simply ground up, providing a poor quality of paper because all of the parts of wood were used, not just the individual fibers. The first paper mill in America was built in 1690 near Philadelphia. The method used at that period produced one sheet at a time and it remained for a Frenchman named Nicholas Louis Robert to develop a continuous process. His first machine was built in 1799 and was patented in England by the Fourdrinier brothers. The Fourdrinier brothers continued to improve the machine but never received any compensation from it because by the time it was adopted as a machine throughout the world for making paper the Fourdrinier brothers had died. Today the Fourdrinier machine is the primary papermaking machine throughout the world. Later a cylinder type machine was invented by John Dickinson and installed near Philadelphia in 1817. The United States now leads the world in the use of paper and paperboard with a per capita consumption of over 600 pounds per person per year making United States the largest consumer of paper products. There are more than 5,000 plants in the USA, manufacturing and converting paper and paperboard. Canada is the largest manufacturer of paper pulp.

The manufacture of one ton of paper takes a tremendous amount of water. About 55,000 gallons of water must be used, most of which is recycled and kept in a continuous process without dumping into the environment. This process also uses sulphur, magnesium hydroxide, lime, salt cake, alkali, starch for binding, some two cords of wood, coal or other sources of energy, alum for improved paper properties, clay (which provides surface smoothness), rosin, dying pigments for coloring and a lot of electricity.

There are basically two types of paper made. The first type is fine paper which is used for writing, printing and some wrapping. The second type is coarse paper, which is mostly used in packaging or for protecting devices. We consider coarse papers to be packaging papers. Since we use almost all wood in our paper manufacture we will take a look at wood composition. When you identify what wood is really made up of, you will find that 50% of it is cellulose. Cellulose is the carbohydrate constituent of the walls and skeleton of vegetable cells. Cellulosic fibers are the backbone of paper. If the fibers come from softwood which is commonly known as evergreen trees or Christmas type trees, these softwood fibers can vary in length from two to four millimeters or close to three sixteenths of an inch in length. Softwood fibers are the longest fibers commonly used and therefore make the best packaging paper because their length provides good strength. Papermaking also uses some hardwoods which are the trees associated with falling leaves in the fall. Hardwood fibers are much shorter, ranging from one-half to one millimeter in length or about one sixteenth of an inch. Hardwood is very abundant in this country and the fact that it has short fibers does provide a certain amount of stiffness to paper. Cellulosic fibers are the backbone fiber of paper. The next composition of wood is lignin. Some 30% of wood is lignin. Lignin is an extremely complicated material. It is dissolved out of the wood in the papermaking process, dried and used as a fuel to help operate the paper manufacturing plant. Lignin's purpose in the wood is as a fiber binder or a glue. The problem that occurs by leaving lignin in paper is that ultraviolet light of sunshine will cause it to turn yellow thus turning the paper to a yellowish color. Since lignin does nothing good for paper it is desirable to remove it from paper. The remaining portion of wood is made up of carbohydrates, protein, resins and fats that take up 20% of wood. Since they provide no benefit to the paper from a strength or surface point of view, they are removed, dried and either burned or in the case of resins they can be converted into other usable chemicals.

To make paper we start with either softwood or hardwood logs. The first thing that is done is that the wood is debarked. The bark is used as fuel. The wood is subject to either one of two pulping processes. The cheapest way to separate the fibers is simply to grind up the wood by forcing the wood logs against tremendously large grindstones that are in a water bath. The water bath cools the grinding stone which keeps the stone from cracking or fracturing and, at the same time, the water will carry off the wood fibers as they are separated from the logs. In the grinding process everything is used and since the lignin and the carbohydrates are still in there the quality of paper is reduced. The grinding process is not used very much for packaging papers. The most used process is more expensive and is called the chemical pulping process. The wood is first chipped into small pieces of about five eighths to one eighth inch thick. Wood fibers are extracted through chemical processes and the nonusable portions of the wood are eliminated. Chemical pulping is a more expensive process but it does provide the best quality paper and the strongest paper. At this point the pulp is washed to remove all the chemicals using a variety of washing methods. All of these chemicals are recycled. The wood fibers are still intermeshed and have some mechanical bonding to each other. The fibers are subjected to a beating or a refining process, with a lot of water added making a slurry called paper pulp which is 96% water and only 4% solids. The purpose of the refining operation is to separate the fibers. Pulp refining can increase the burst strength and tensile strength of the paper. Refining is a time controlled situation so that the manufacturer of the paper can provide the customer with the properties that they so desire. Also during this process of refining and beating, sizing materials are added. Sizing materials are very important in the paper manufacturing. Sizing adds such things as starch to increase the bonding of the fibers. Resins can be added for increased water resistance of the fibers. The addition of pigments and clay will add certain properties to the paper such as ink hold out, coloring, stiffness, and water resistance. There are many sizing materials that can be added at this time to satisfy the needs of the consumer.

Once the pulp is properly prepared it then goes to one of two machines. The primary paper making machine is the Fourdrinier. Modern paper making machines are huge. They can be as long as a city block and several stories high. They produce paper up to 30 feet wide and usually operate around the clock at speeds of 3,000 feet per minute or 800 miles of paper per day. These can be tremendously varied as far as width or speed is concerned. The Fourdrinier machine was developed in England and has been refined and improved throughout the years. It is the primary papermaking machinery used throughout the world. The Fourdrinier is used to make mostly thin paper that is up to 12 points in thickness. Thicker papers can be made on this machine if needed. Most Fourdrinier machines make only one layer of material but it's possible to add another head box to these machines and put down a second or even a third layer of pulp to the web. The slurry flows onto a continuously moving copper wire screen that moves in a racetrack like pattern. During this flow onto the copper screen the screen is shaken from side to side to reduce fiber alignment and allow more fiber interplay or mechanical bonding to the fibers. This action will result in less fiber alignment and make the paper stronger. The pulping water is continuously removed first by gravity, then by suction and then by pressing to the point that, by the time the paper reaches the end where the screen doubles back in a racetrack like pattern, the paper is strong enough to support itself and go into the pressing and drying sections. The pressing section simply irons out more water by pressure. The paper is then fed around a series of drying rolls that are steam heated. When the paper gets to the far end of the papermaking machine it has gone from 96% moisture down to around 5% moisture and is considered dry at that point. The last function of the paper machine is calendaring which is a stack of highly polished rolls through which the paper is fed making the paper smoother. The paper is rolled up at the far end of the machine. Then it is immediately removed and another roll is started without stopping the machine. The completed roll of material can then be taken to other machines for cutting to the proper width, for laminating, for coatings as needed or any other modification to convert it into usable forms.

The second most used type of papermaking machine is called the cylinder machine. The cylinder machine makes heavy grades of paperboard. It uses anywhere from six to eight separate vats of pulp. The vats of pulp can have varying mixtures of pulp. Recycled paper pulp is used extensively in cylinder machines. It makes paperboard by building up layers from each of the vats of pulp. Cylinders roll in the vats and deposit layers of pulp on the underside of a felt blanket. After building up several layers of paper pulp on the felt blanket, the paperboard leaves the felt and goes into a pressing and drying operation much like a Fourdrinier. The felt also goes through a pressing and drying operation and back to pick up more paper from the cylinders to complete its continuous cycle. The wet paperboard continues on its own to the drying operation. Paperboard is much thicker than paper, so the drying operation is far more extensive reaching as high as three or four stories with multiple rolls of steam heated cylinders that drive the moisture out of the paper. The paperboard works it's way up and then down through these drying cylinders until it reaches the proper moisture level at which time it can be coated. Coatings provide a nice white printing surface. The paperboard is next rolled up, removed from the papermaking machine when the roll is complete. A new roll can be started immediately without shutting down theprocess. The rolls of paper are taken to another area where they are cut to size for the next operation. One significant problem from the cylinder machine is that it does have heavy fiber alignment as there is no means to shake screens and cause fibers to overlap. Paperboard has a lot of fiber alignment in it which the packager has to be aware of. The big advantage of the cylinder machine is that it does use a tremendous amount of recycled paper and it can build up thick layers to gain strength. Typically paperboard varies from 12 points up as high as 70 points in thickness although it can go even higher.

All paper is bought and sold on the basis of weight or basis weight. Basis weight by definition, is the weight in pounds of a ream of paper. A ream of paper is equal to 3,000 square feet of surface. The basis weight of paper is pounds per 3,000 square feet. The basis weight for paperboard is pounds per 1,000 square feet of paper. When ordering paper for use in packaging all the packaging specification has to say is that we desire 50 pound paper of such and such a thickness and the manufacturer can provide this specifically. The basis weight of paper is controlled by the amount of pulp flowing onto the screen or being transferred from the vats, and the speed of the pick up screen or felt as it traverses through the machine.

Thickness of paper can be controlled by calendering, pressing and the amount of fiber that is put onto the screens. If you ordered 500,000 reams of 20 pound paper, you can easily compute out the total weight of a shipment of paper. In this case it's 20 pounds times 500,000 reams which equals 10 million pounds of paper. You can also compute the surface area of paper that you're going to get knowing that there is 3,000 square feet per ream, and if you have 500,000 reams of paper, multiply that by 3,000 square feet of paper per ream which will give you the surface area of the paper.

If we look at paper as an industry certainly the largest use of paper is for printing, writing, wrapping and sanitary purposes. Some 29 million tons of paper are manufactured each year in this country and of that almost 20 million tons are used for fine papers for printing and writing. About five and a half million tons are used in packaging. The remaining four million tons are usually in the form of tissue paper that's used for a variety of uses. The strongest and most popular paper used in packaging is called unbleached kraft. It is a coarse brown paper produced on the Fourdrinier machine at very high speeds. It comes from softwood pulp almost always made by the chemical process and it is the most economical and strongest packaging paper available. A good example of this paper would be a grocer's bag or typical sack paper. Kraft paper is generally characterized by a rough finish and the basis weight varies from 20 pounds to 60 pounds. Kraft paper is easily modified into many shades of brown and acceptable for surface printing. It is used extensively for bags, also for envelopes, sealing tapes, laminated papers, multiwall paper sacks, wax papers and so forth. Kraft paper can be bleached. Bleaching removes the browness of the natural paper. Bleaching papers reduces overall strength but it makes a better printing surface and if calendared extensively it can be an excellent printing surface. It is commonly used for overwrapping or laminating of boxes used for bags and surgical packaging and is still very machineable. Another paper that's commonly used in packaging is pouch paper. Pouch paper is made from bleached virgin kraft paper. During the manufacturing process the paper is super calendared to be very smooth and then treated chemically to be pliable and to have a semi-gloss look. It has great strength and is commonly used in food packaging especially making pouches or bags for food packaging.

Another paper that is commonly used to protect products especially for bagging products that are in folding cartons is called glassine or greaseproof paper. These are very dense papers made from hydrated pulp that are super calendared for smoothness resulting in a very smooth, dense, semi-transparent paper. During manufacturing the pulp is highly beaten to break down fibers and get a high degree of hydration which is important. The properties of this paper are grease resistance and a high resistance to air passage and oil vapors. Glassine is a very good food protection wrapper, especially for butters, margarines and all kinds of food stuffs, plus tobacco and metal parts. The paper is made into bags, box liners and wrappers, and is extensively used for candy bars, crackers and safety seals on top of glass jars that contain products that are moisture sensitive such as instant coffee and cocoas. The normal color of glassine is whiteish but it can have various colors. Plain glassine and greaseproof paper is not water resistant. They will allow moisture to penetrate through them unless they are waxed or lacquered. If waxed or lacquered they can resist moisture and moisture vapor penetration. Common usage in the home is as waxed paper. Another paper that is used in some packaging is tissue. Tissue is a very lightweight paper, usually less than 18 pounds basis weight. Tissue is relatively smooth and it can have a soft finish or a hard finish. Tissue is usually a very pliable and somewhat translucent paper mostly used as an inner wrapper for protection of dry goods, wrapping of flowers and twist wraps for candies and other foods. Tissue papers can be waxed for better moisture resistance.

Cellophane is also a paper. Cellophane is a very transparent film made from regenerated cellulose. It is grease proof and can be made moisture proof and heat sealable by suitable coatings. Cellophane was the first transparent flexible material that was available for packaging and for years was used for overwraps and numerous other wrapping material because of its transparent property. Cellophane is not a thermoplastic, nor is it heat sealable. It is very sensitive to moisture, it does have good strength but it has very poor tear resistance and tends to dry out and tear easily when sitting on the shelf. From a marketing point of view it has excellent clarity and sparkle. From a packaging point of view it's easy to machine and makes a strong package. If notched at the proper point, it is easy for the consumer to open and remove. Cellophane is expensive when compared to polypropylene which has very similar properties at less cost. It also has a short shelf life because of the drying out effect and it tends to shrink and become brittle when dried. In recent years the demand for cellophane has significantly decreased because of cost and aging properties that are not a problem in some of the plastics.

Paperboard is the most used type of paper in the packaging industry. It is a heavyweight thick sheet of paper mostly 12 to 30 points in thickness and can be as great as 60 points for some packaging. The biggest use of paperboard is in the manufacture of corrugated cartons which are the primary shipping container used throughout the world. The second use of paperboard is for folding and set up cartons for the packaging of consumer and commercial goods. It can be made to have an excellent printing surface. Paperboard is known for its excellent machining properties. It has good shelving because of its rectangular shape. It is a low cost product that is used extensively for packaging quality products that do not require a long shelf life or extensive protection. There are several types of paperboard that are used. The cylinder papermaking machine makes chipboard which is made from 100% recycled paper. It's a very inexpensive material. Chipboard is not good for printing because it has the color of the recycled paper from which it is made. Chipboard is mostly used for the manufacture of set up boxes, such as shoe boxes, candy boxes, stationary boxes and gift boxes. Cylinder machines also make white lined chipboard. This is a recycled paperboard product with a white lining that provides excellent printing and it is used to manufacture folding cartons that can contain food items that are first put in a liner bag of plastic or glassine. Typical of white lined chipboard use is for dry cereal boxes and other boxes. The Fourdrinier machine makes clay coated natural kraft. This is a very strong paperboard. The clay coating provides a very printable surface and because it is natural kraft paper many sizing materials can be added. It is made to be moisture resistant and is commonly used as beverage carriers such as for six packs. Another paperboard obtained from the Fourdrinier machine is uncoated solid bleached sulphate paperboard commonly called SBS. SBS is 100% virgin bleached pulp that provides a strong white water resistant paper mostly made into folding cartons to contain frozen foods. This type of paper does not require a liner since it is made of virgin bleached pulp. There are many other uses of this type of paperboard which makes it very popular in the packaging industry.

The largest use of paperboard in packaging is to make corrugated board. Corrugated board is defined as a packaging material consisting of a central member which has been fluted on a corrugator onto which two flat sheets of paperboard have been glued making a three layered structure, flat on both sides and corrugated in the middle. Corrugated board is much stronger than the material of which it is made due to two architectural structures that are the arch and the column. The arch provides strength in thickness and the column provides strength in stacking. By combining these structures to make corrugated board we have a very strong rigid lightweight container that can provide some cushioning effect for the products and insulation protection for the products. Corrugated board is made by passing the corrugation material through a machine that first softens it by applying a spray of steam to the paper and then running it between meshing gears which causes the corrugation of the paper. The corrugating rolls are heated to drive out the moisture originally applied so that the board will stay in the fluted or corrugated shape. Also prior to leaving the corrugation rolls one side of the corrugated paper has glue applied. Linerboard is glued to the corrugated board as it comes out of the corrugator resulting in two of the layers being assembled. The two combined layers then receive a second layer of linerboard which makes the complete corrugated board which is stiff and from that point on must be cut into the size of blanks that are required to make the appropriate containers. The corrugation that has been formed is called flutes and in this country we have four different flute sizes. The first flute made is identified by the letter "A". "A" flute has 36 flutes per linear foot and is about three sixteenths of an inch in height. As corrugated board began to be used during the early part of the 20th century users found that this corrugation was too large. Users requested a smaller corrugation which was designated as "B" flute and contained 50 corrugations per linear foot and was only three thirty seconds of an inch in height or one half the height of "A" flute. "B" flute turned out to be too extreme a change. The common size used today in most manufacture or corrugated board is known as "C" flute which has 42 flutes per linear foot, nine sixty-fourths of an inch in height. In recent years there has been an increasing demand for what's known as "E" flute. "E" flute is a very small flute that has 94 flutes per linear foot and is only three sixty-fourths of an inch thick, making it slightly thicker than common paperboard. The purpose of "E" flute is to provide a strong rigid package that is lightweight. Because of the number of flutes per linear foot it has a much smoother printing surface and it is commonly used today for primary packages of small appliances such as irons and toasters. "D" flute was not made.

Construction of corrugated board comes in several ways. It is sold as single faced, which is a facing with a wall—only two layers. An advantage of single faced is that it's a very flexible material and is used mostly as a cushioning and decorative material. Light bulbs are commonly packaged in single face corrugated board. The most common structure is single wall or double faced which has three layers—the two flat outside layers with a corrugated layer inside. These are used extensively for shipping containers, liners, pads, and partitions. This is the most used corrugated form in packaging. If greater stacking strength is needed for heavier items, double wall corrugated board which has three facings and two corrugations provides a tremendous increase in strength for packaging televisions, computers and other things that need additional protection. Triple-wall corrugated can be used for the very large products like refrigerators and stoves. It has greater strength without the weight of solid wood boxes or crates. Triple wall corrugated can be used alone or with wood for packaging of large heavy articles.

Corrugated case manufacturing is the largest single packaging industry in United States. The use of corrugated shippers had a slow start at the turn of the 20th century. Corrugated had to compete with wooden boxes which were required for shipment by the railroads. The trucking industry was trying to get started and they would accept corrugated shipping cases which forced railroads to accept them also. Now the corrugated box is the preferred shipping container througout the United States. Sales of corrugated follows the USA economy and is used by many economists as a fast indicator of what the economy is doing. Corrugated sales parallel the GNP. Food packaging or shipping is the largest user. Twenty-five percent of all corrugated is used as a tertiary package of foods. There are about 1,400 converting factories in the United States. Most cities of any size have corrugated converting plants which buy roll stock from the paper mills and convert it into corrugated materials complete with printing and cutting, slotting, and gluing ready for the packager to package and ship products. Today 98% of all packaged US produced products are shipped in corrugated containers. Consumption over the past few years has increased at 3% per year and it is estimated that the United States will soon exceed its corrugated industry capacity and the packaging industry will probably experience greater price increases in corrugated board.

Section D: *Plastics in Packaging*

Plastic is a material capable of being formed in various shapes by heat or pressure or time. There are two types of plastics used today. The first plastic is called thermoplastic which are plastics that react to heat by softening. Thermoplastics can be repeatedly heated and reformed. Almost all packaging plastics are thermoplastic. The other type of plastic is called thermoset. When thermosets are originally heated they permanently set and cannot be reformed. Thermosets cannot be recycled.

All plastics are made from crude oil, natural gas and organic chemicals. They have a carbon base and can be burned to recapture the energy used to make them. Plastics are composed of chain-like molecules of high molecular weight called polymers. The terms polymers and plastics are synonymous. A monomer or combination of monomers are used to manufacture plastics. A monomer is a chemical compound that can undergo polymerization. Different types of monomers or a combination of different types are used to make different plastics. Plastics manufacturing starts with a single molecule that attaches to other like molecules to form a chain. The term polymerization means many parts together. Polymerization is a chemical reaction where two or more small molecules or monomers combine to form large molecules. They contain repeating structural units called chains of the original molecules. A polymer is defined as a compound formed by the linking of simple and identical molecules that have higher molecular weight.

The first plastic known was found in 1843 when a Frenchman in Malaysia discovered natives using a natural material called gutta-percha. It is a rubber like sap from a tree that has a heavy resin. The Malaysians were forming this gutta-percha into knife handles and some of this material is still being used today in combination with other materials to form simple things like buttons, checkers and personal articles. In 1845 a chemist was given the task of finding a substitute for ivory billard balls. He developed cellulose nitrate in his laboratory. This was the first commercial man-made plastic. In 1933 polyethylene was discovered as a result of waste from another chemical reaction. Polyethylene's first use was in World War II as a wrap for radar cable that was being submerged under water. It wasn't until the 1950's that the use of plastic became popular in packaging. Now about 25% or greater of all plastics manufactured in the United States are used in packaging. Of that 25% percent more than 47% of it is used for the manufacture of packaging films. Forty percent of packaging plastics are used to make containers, 6% for closures and 5% for adhesives.

Plastic enjoys many advantages over other packaging materials. One advantage is the fact that plastics are lightweight which is a big factor when shipping packaged goods. Plastics have a high weight to strength ratio. Most plastics have an excellent resistance to products. Plastics have a good ability to be tailored to the needs of the product. Plastics shape very well and are easily modified. Plastics can be soft or they can be rigid. They can be very thin or they can be very thick. Thin walls of plastic containers take up less shelf space. Plastic forming machines used to blow mold package containers are very small and can easily fit into the packaging production line to make plastic containers just prior to filling thus reducing storage space needed for raw materials and preformed packages like empty cans and empty bottles.

Packaging plastics can be made into either films or sheets. The big difference between the two is flexibility. Plastic films are much thinner, generally less than 10 mils in thickness. Plastic sheets are greater than 10 mils in thickness, less flexible, sometimes less transparent and used primarily to make other formed containers such as blister packs or skin packaging. Packaging plastics can protect products from mechanical hazards associated with handling, transportation and stacking, climatic hazards such as rain, snow, temperature extremes, light and moisture, biological hazards such as insects, bacteria, mold, rodents and birds, and social hazards such as pilferage.

Currently there are about 12 different types of plastic films used in packaging and these films are mostly used in combination with each other or other packaging materials to obtain the property desired. There are known to be well over 7,000 different combinations of films on the market today but they all come from these 12 films. Of these 12 plastics there are only 4 that have most of the packaging market. The plastic most used is called polyethylene, or the common term used throughout the industry is PE. PE film is used for most everything and the most commonly used polyethylene is called low density polyethylene (LDPE). Polyethylene is quite translucent, so it can be used to show products. LDPE tends to be a very limp film, which is difficult to machine. LDPE attributes are good flexibility and good moisture protection. It also is one of the lowest cost films used in packaging. It has good low temperature capability, maintaining good properties even when very cold. Low Density Polyethylene has a high gas permeability and high migration of odors. It is not a good barrier for carbon dioxide or oxygen. LDPE is not good for rigid containers because of its softness. It is difficult to open LDPE packages because it tends to stretch and tear poorly. The properties associated with low density polyethylene and the fact that it's not a good oxygen or a good gas barrier come from the fact that the chains tend to be highly branched. LDPE can be compared to a brush pile that has lots of holes in it providing lots of air space for gases to migrate through. LDPE has excellent moisture protection. To improve the properties of polyethylene a chemical process was developed which reduced the branching effect on the polyethylene chains. This lack of branching produced a polyethylene that was a better barrier material but it lost clarity. This modified PE is called high density polyethylene or HDPE. High density polyethylene is primarily used for blow molding bottles. HDPE is slightly higher in cost than low density polyethylene but is still an economic plastic. There are other polyethylenes on the market today providing various changes in properties and they're named medium density polyethylene and others. The polyethylene family of plastics starts from the polymerization of ethylene gas under pressure. Ethylene is a combination of hydrogen and carbon.

Another plastic that is extensively used in packaging is polypropylene or PP. PP comes from the combination of propylene gas with carbon and hydrogen where a propylene molecule is added to the chain in lieu of one of the hydrogens. This combination makes a film that is very high weight compared to others but has excellent clarity and sparkle. PP is commonly used as a replacement for cellophane which is more expensive and has shelf stability problems. PP has a high melting point which makes it good for ovenable dishes and for lamination onto other plastics to form good protective capabilities such as the squeezeable ketchup bottle. PE is used to make closures of all types and is used to package intravenous solutions in hospitals. From a cost point of view it's slightly higher than polyethylene, about 20 to 25 percent greater in cost.

Another plastic that has a great share of the packaging market is polyester, which is abbreviated to PET. PET is a combination of ethylene glycol plus terephthalic acid. Polyester is very complex molecule once polymerized, provides great tensile strength and good high temperature resistance. PET's primary use in packaging is for carbonated beverage bottles and other types of plastic bottles. PET is only a fair oxygen barrier and therefore limited to the types of beverages that can be packaged in it. Carbon dioxide still will leak out of a PET container but the typical one litre bottle of carbonated beverage will have a shelf life of about eight weeks. PET is an expensive polymer costing almost three times more than the most inexpensive polyethylene. PET has extensive use outside of the packaging field for clothing, computer tape, recording tape and other things.

The fourth popular packaging polymer is polyvinyl chloride or PVC. PVC is one of the older, original plastics formed by a combination of ethylene and hydrochloric acid. A chlorine atom is substituted for a hydrogen atom. PVC is a more expensive film, slightly greater in cost than is polyethylene and more costly than polypropylene. PVC is a very tough film that has great clarity. PVC is commonly used for packaging of oils and alcohol solvents. The mechanical properties of PVC are that it is a very limp film, it tends to cling and it's very difficult to machine. PVC is often laminated to other materials to enhance machining capability. PVC can be blow molded into bottles that have a good squeezing capability and are commonly used for shampoos, cosmetics, household chemicals, cooking oils and others. Both PVC and polyethylene as films are used extensively for stretch wrapping pallet loads of packaged goods. Nonpackaging use of PVC is to make plumbing pipes and phonograph records.

There are many plastics that can be foamed for cushioning materials. One of the well known foamed plastics is polystyrene, which is abbreviated as PS and manufactured by combining ethylene gas and benzyne. A benzyne ring is attached to one of the carbon atoms in the molecular structure. A common use of foamed PS is for egg cartons. The "peanuts" that are poured around products in a package to provide protection against shipping are commonly expanded PS. Expanded PS is manufactured by steam heating little beads of polystyrene that have been saturated with air. The heating of PS beads causes a softening of the polymer and an expansion of the air to give a spongy like property. PS beads that are heated and expanded within a mold will conform to the shape of the mold thus making egg cartons or coffee cups. Some clear bottles are made from polystyrene and used for vitamins and spices. PS is used in a nonpackaging market for several things. One of the most common uses of PS is to make toilet seats.

The forming of plastic parts such as closures and/or bottles uses a system called injection molding. Injection molding uses plastic resins that are continuously flowed into a torpedo like machine that forces resin through a heating chamber to melt the plastic and force it into a mold or through a dye to get various parts. Injection molding can be a continuous process or a stop start process. The same melting procedure is used to make plastic films by forcing melted plastic through a die that has a very thin slit forcing the plastic out in sheet or film form. Plastic is also blown in a large bubble and then collapsed to make films. The blow molding of plastic containers is much like making glass containers. Sometimes plastic bottles are formed in two stages, first forming a parison or premature shape and then the final shape. Two stage forming is commonly done in polyester or PET containers to improve control of sidewall thickness and to gain strength.

Thermoforming of plastic containers uses plastic sheets which are heated to the point that they start to drape or stretch and then allowed to come in contact with either a male or a female mold and with the assistance of a vaccum draw that will pull the soft plastic material onto the molds forming blisters or cups.

In summary the overall advantages of plastics compared to other packaging materials is a lightness in weight and also that plastics are very strong compared to their weight. Packaging plastics can be classified as inert, having good resistance to almost all products. Plastics can be tailored to meet the needs of a product, having great flexibility in forming and this forming can be done right at the time of packaging or as a filling operation is taking place. The thin walls of plastic containers allows more product to be stacked into and onto the shelves, railroad cars and trucks. Plastic forming machines tend to be very small and fit nicely into the packaging line. Products packaged in plastic bottles don't require partitions between packages to reduce breakage. Packaging plastics do have several disadvantages. One disadvantage of plastics is the overall lower strength when compared to glass and metals. Plastics have stability problems of warping, shrinking, and creeping. Most plastics have low thermal limits. Most plastics will burn and some will give off toxic gases during burning. There are some very good plastics that cannot be marketed because they have a very strong objectionable odor and would not be acceptable in the industrial and consumer environment. Cost of plastics is also a disadvantage because cost fluctuates with the availability of hydrocarbons and cost of plastic is competitive only when used at large quantities. The squeezable ketchup bottle is far more expensive than the equivalent in glass but the American consumer has accepted this higher cost by gaining the squeeze and nonbreaking capabilities that are not possible in glass. The future use of plastics in packaging appears to be growthy as more and more engineering is developed to combine materials that provide better protection. Plastics are expected to find more markets within the packaging field and is considered to be the only material used in packaging that has a bright future.

Section E: *Wood in Packaging*

Wood has been used in packaging for hundreds of years and most packagers consider the wooden box, crates and barrels to be the first type of modern shipping containers. There were two reasons that wood enjoyed a lot of use in this country. First, lumber was very plentiful and still is to the extent that a lot of lumber in the United States is exported. Secondly, lumber was inexpensive up until recent years. These two facts attributed to the popularity of use of wood in packaging. Now the cost of lumber and the cost of labor in converting lumber into packages such as barrels and crates make wood an expensive packaging material. In the United States, containers were made by the packagers in a carpenter shop to fit the product and provide protection to this product throughout the distribution system. Most wooden containers are now made by suppliers and shipped to the users in a flattened condition. The user erects the wooden containers, fills them, seals them and ships them out. Competition from other packaging materials such as corrugated board has signficantly reduced the use of wood. Wood is still used for superior protection in shipping and distribution, mostly for industrial packaging in the form of crates and boxes. There are a few luxury packages made of wood for things like jewelry, special promotion items, some instruments, cigars, and wine cases. The biggest use of wood in packaging is for pallets. A pallet is a platform designed for easy use with fork lift trucks. A pallet can accommodate several shipping containers to make a unit type of package for distribution.

The basis of wood costs and strength come from four factors. The first is the type of wood. There are very expensive woods like walnut and mahogany and there are less expensive woods like oak and pine. The second factor of wood cost is the quality of the wood. Quality of wood is dependent upon the grain, defects, stains, cracks, and checks. The third factor of wood cost is the thickness of the wood to be used. The fourth factor of wood cost is the workmanship. You can buy rough cut woods for pallets which are low in workmanship or you can get smooth woods which have extensive machining and cutting to make high quality surfaces. All woods come from two categories: soft woods and hard woods. The differences of these woods was covered in the paper section, but for construction purposes in packaging, softwoods are mostly used to make crates and boxes. Softwoods are easy to cut and work. Outside of packaging the typical use for softwood would be in house and building construction. The second type of wood is hardwood. It is used in packaging as pallets and loading platforms. Outside of packaging most hard woods are used for furniture and flooring. The strength of wood related to moisture content. When wood has high moisture content strength will be lowered. The lower the moisture in wood the higher the strength will be. Green wood, which is freshly cut wood has only 30% of its potential strength. At 12% moisture (normal dried wood) the strength of the wood will be doubled. Wood that is dried down to zero percent moisture will triple it's strength but 0% cannot be maintained outside of drying ovens. One of the problems of using wet woods is that they have the tendency to warp and/or split as they equilibrate thus even further reducing their desirability and strength. Wood fasteners that are used in construction are a big factor in the strength and durability of the container.

Wood fasteners are primarily nails, screws and sometimes nails and screws in conjunction with strapping. Nails for wood construction are of several types. The cheapest type of nail is common bight nails. Nails can be cement coated which increase their holding power by 40%. Nails can have a thread that will increase their holding strength 250% but the cost of the nail will increase. Use of screws in wooden container construction is very expensive because it is a labor intensive process. Wood screws are mostly used for recyclable or reusable boxes because they provide a tremendous amount of strength and durability that allow the boxes to be used over and over again. Strapping is commonly used to strengthen both nails and screws around strategic points of wooden shipping containers. Strapping can be steel or plastic that improve the structural integrity of the container. Nails come in multiple sizes. Nail sizes are suggested by the U.S. Forestry Products Division in Madison, Wisconsin. There are specific standards for types and sizes of nails for various types of woods and various sizes and types of shipping containers. Nails are sized by the term "penny" which with a number denotes a certain diameter and length of nail (i.e. 6 penny).

Specific wooden containers can be outlined as a crate, which is a structure in which the frame members sustain the load and define the shape. Crates can be purchased in all sizes from apple crates to large machine holding crates and they can be either open or closed up tight. Unsheathed or open crates are less expensive but they provide little protection except for stacking and bumping from other adjacent containers. Wooden boxes tend to be closed up completely. They are made of dimensional lumber, and can be purchased in several styles that are published by the Forestry Department of the Federal Government. Wooden boxes can be of simple boards nailed perpendicular into boards to form the package or they can contain cleats which will significantly increase the strength and durability of the container.

Water type or fluid type containers are generally called barrels. Barrels are cylindrical in shape and usually made out of the hardwoods that are suitable for containing products. Wooden barrels are usually bilged, which means the diameter of the center portion is larger than either end. Bilging facilitates the tightness of a barrel and also makes the barrel easier to roll. One person can handle a bilged barrel by rolling it throughout the warehouse and into and out of trucks, and when it is desired to upright the barrel that person can start a rocking motion and tip it upright. Barrels come in various sizes, everything from small one gallon size up to sixty gallons. Barrels can be either tight or slack. Slack barrels are not suitable for containing liquids but can be used for powders, semi-solids and solids and when properly lined, can carry semi-liquids. The tight barrels are manufactured to be liquid tight and can be used for any product.

Wire bound boxes and crates enjoy popularity within industrial packaging as shippers of large and bulky items. They are wooden frames with veneer slats that are fastened by steel wire. Wirebound boxes are a lighter and open to air container. Their strength and impact resistance is satisfactory for many applications and the fact that they use veneer or resawed lumber reduces significantly the weight of the containers. Wirebound boxes are purchased and shipped knocked down flat, then erected by the user.

There are over 340 plants making wood containers for the packaging industry. The manufacturer and use of pallets is the biggest user of lumber. Pallet loads can be shrink or stretch wrapped or strapped and can be handled and moved as a single unit. Current sales in pallets each year in this country exceeds one billion dollars and is the largest single wooden industry in packaging having over 1,600 producers. Producers of pallets tend to be located in the areas of industry where they are going to be used.

Lumber is measured by board feet and when purchasing large quantities of lumber it is ordered by the board foot. By definition one board foot is equal to a piece of wood that is one foot long, one foot wide and one inch thick. When computing board footage you must divide the length, width and thickness by 12 so that it fits the definition of a board foot. For instance if you have a two by four by twelve foot long board you divide that by twelve and you will get eight board feet. If you have many boards of the same dimension you multiply the whole equation by the number of boards to determine total board footage. If you want the total price of this lumber then you multiply that equation times the price per board foot.

The use of wood in packaging is only 3.3% of the total packaging materials, but wood is a very essential part of the packaging industry. There are many other materials used in conversion and manufacturing of packaging systems that are too numerous to name. They include things like labels, tapes, twines, cushioning, adhesives, coatings, inks, staples, and others.

CHAPTER 3

Flexible Packaging

Flexible packaging is the oldest form of packaging known. It originally used any suitable material that was available. Glass and leaves were used and from simple materials, mats and cloth could be made for wrapping. Simple, natural flexible materials for packaging have been used for many years and in some parts of the world they are still used. In Southeast Asia it is common to buy products from open markets that will be wrapped in banana leaves. All suitable materials have been used throughout the years but in today's market there are three materials that are used. Today's flexible packaging uses paper, plastic, and metal foils or combinations of these three materials. Current sales of flexible packaging is close to 10 billion dollars a year and is forecast to have rapid growth. Flexible packaging provides an economical, versatile and lightweight package in the form of wraps, bags and pouches, most of which are single web. But when they are combined with other materials such as laminates, a significant increase in the properties of the materials is gained and their value to the packaging industry is enhanced. The standby for flexible packaging is paper.

Paper is very popular with packagers because it has excellent machinability, excellent folding properties, excellent printing and is the lowest cost material available. Flexible paper packages are commonly used for cereal packages, butters, used as overwraps, numerous types of bags from tea bags to pet foods to flour, seeds, fertilizer and others.

Plastics are used because of their transparency, heat sealing ability, and moisture resistance. We use polymers such as polyethylene, polypropylene, polyester and polyvinyl chloride and many others.

Aluminum foil is the primary flexible metal that is used. Aluminum foil provides the best barrier material that is available.

Aluminum foil provides an attractiveness to the package by its color and reflectivity.

The various wraps that are used as flexible packaging are used for many reasons. One of the uses is to insulate individual or unit products from the others like sticks of butter or sticks of gum. They are used to cover primary packages as an overwrap such as cigarette packages. They are used to bundle or unitize containers that have two or more products in it such as the variety pack of individual cereal containers. They are used to shrink wrap both primary packages and/or pallet loads of tertiary containers where they excel in holding and compressing products. The primary shrink materials are polyethylene and polyvinyl chloride. During the manufacturing process of shrink films, the films have a memory built into them by stretching while in the soft stage, then cooled. This memory allows the material to shrink back to its original size when subjected to heat. Flexible wraps are also used in a stretch form. Stretch films have the same function as shrink films but the cost is significantly less because there is no heat energy used. Stretch films are manufactured to have an elasticity and these films are mostly polyethylene and polyvinyl chloride.

Bags and envelopes that are made from flexible materials have many construction styles from the self opening square bags commonly found in grocery stores to satchel bottoms that are used for flour and other products. Strength of these bags can be significantly increased by making multi-wall bags from the combination of many layers of paper and/or plastic. Some bags can be made insect repellent through suitable coatings. Envelopes are very inexpensive packages. They can use materials like glassine to kraft, paperboard to cellophane, plastic to foil. Envelopes can package pills to turkeys. A recently developed laminated flexible package is the retort pouch which is sometimes known as a flexible can. The retort pouch is a laminate of polyester, foil and polypropylene. The aluminum foil is buried between polymers for protection from damage as it provides two functions that are the barrier capability of the pouch and the aesthetic appeal of the foil. Retort pouches are filled and sealed and then sterilized at 250 degrees fahrenheit. The retort pouch is a flat pouch. When sterilized or retorted the product does not get overcooked but all of the product gets equally cooked resulting in a product quality that is better in taste than those that have been packaged in metal cans. The quality of retort pouched products is not as good as frozen foods. The advantage of retort pouched products is that it's better than canned products but not as good as frozen. Retort pouches do not require storage in a freezer. Retort pouching has a slow machining speed so the cost of pouching is higher than conventional canning or bottling. When pouching technology catches up to make it more competitive, the pouching of foods could be a primary method of packaging and preserving food items.

A semiflexible package that has been marketed in recent years is called the brik pack or aseptic packaging. Aseptic means the absence or exclusion of microorganisms. The aseptic system combines a pre-sterilized product into a pre-sterilized package under aseptic conditions. The aseptic package is made up of five layers, that are laminates of polyethylene, paper and foil. These five layers have polyethylene on the outside which provides moisture protection to the rest of the package. The next layer is paper which has two functions. Paper gives structure to the package and also provides an excellent printing surface. The paper is laminated to a thin layer of polyethylene to which aluminum foil is laminated. The inner layer of PE has a primary function to laminate the paper and foil together by acting as an adhesive. On the inner side of the foil is another layer of polyethylene whose function is to protect aluminum foil from damage and to provide a heat sealing surface for making the pouch. The aseptic package cost has a favorable advantage of 30 to 50 percent less than metal or glass containers. Studies show that at least six months of shelf life can be obtained through this system without refrigeration. The single serving juice market has mostly gone to aseptic containers and they can be found in both markets and in dispensing machines. The future looks good for flexible packaging. The packaging industry expects that there will be a gradual shifting from rigid containers to flexible containers as protection and machinability are improved.

CHAPTER 4

Aerosol and
Tube Packaging

An aerosol is a valved container of liquid or powder that is under pressure and will discharge the product through the valve on demand. Aerosol dispersed liquid product particles can be so small that they will remain suspended in the air for a long period of time. Other products are dispensed in a manner so that their use is enhanced. Cheese spread is dispensed from an aerosol in a manner that allows it to be spread on surfaces. Powders are dispensed in a suitable concentration that allows them to do the job that they are intended for. Aerosol systems have become essential for some products. Some medicines require a very fine particle size that only an aerosol system can provide. Shaving soaps in aerosols are ready to go to work when dispensed. Other products are simply convenient when dispensed from aerosols because they could be used just as good coming from other types of containers, but the consumer prefers the convenience of the aerosol. Paints, insecticides, deodorants and perfumes are all dispersed from aerosols as a convenience.

The aerosol container was developed in the United States during World War II at the request of the United States Army which was getting more casualties in tropical areas from insect carrying tropical diseases like malaria than from combat. The United States Department of Agriculture developed what was then known as the "bug bomb". They converted an existing steel beer can into a pressurized package that was valved. The "bug bomb" was very successful and a very efficient package for the army to use. In 1947 the first commercial use of an aerosol was realized as an insect aerosol and about five million containers were marketed the first year. In 1953 the aerosol propellants were changed from fluorocarbons to hydrocarbons to reduce cost. Also in 1953 shaving soap which is now a major part of the aerosol container entered the market. Soon other products were packaged in aerosols. Some aerosols were successful and some were not, but in 1974 the aerosol program had a big setback from three problems. First, there was a fluorocarbon theory published by scientists working in California that theorized the fluorocarbons released through the aerosol system were attacking and destroying the ozone layer above the earth. It's important to realize that, at this time, this was a theory and was not proven. Recent studies have supported the theory and it is now accepted by the scientific world that released fluorocarbons are causing ozone problems. The second problem that occurred at this time was a recession. During a recession money becomes tight and some of the first things that consumers start giving up are luxury items. The aerosol container can often be classified as a luxury package. It is a very expensive package that is working for the consumer. Products can be obtained in nonaerosol packages at a lower cost. The third problem was caused by poor quality control. The aerosol containers had a series of explosions and/or leaks which caused consumers to avoid some aerosols.

Current sales of aerosols are two and one-half billion dollars per year. Of that two and a half billion, thirty-three percent of total sales go for personal items like hair spray, deodorants, shaving soaps and perfumes. Twenty-five percent of sales go for household items like cleaners, room deodorants, disinfectants and others. Fourteen percent of the aerosol market goes to the industrial automotive sector in the form of lubricants, refrigerants, paints and others. Thirteen percent of the market goes to coatings and finishes that are consumer paints, varnishes, and others. Eight percent of the aerosol market is for insect spray and the remainder goes for food and animal care products and miscellaneous objects of varying substances.

Liquid products contained in an aerosol are released by pressure through a push-button valve. It is estimated that a one second discharge will send out into the atmosphere one times ten to the eighth droplets. The aerosol system consists of a container, a metering valve, a dip tube, the product and the propellant. The volume size of aerosols sold in this country vary from just a few grams that are used in drugs and perfumes up to 24 ounces that are used for thinning fluids and other liquids. An aerosol is an expensive system, costing about $1.30 for the can with the valves, and the propellant. The overall cost for packaging an aerosol product is about four times greater than packaging by conventional means.

Aerosol cans are mostly metals. Steel is the primary aerosol can metal taking about 74% of the market. Aluminum cans take about 20% of the aerosol market. Aluminum cans tend to be smaller than steel cans. Some stainless steel aerosol containers are used and are usually of a large size that are made to be refilled and reused. There are a few glass containers, less than 5%, that are marketed for their looks and their inertness. They are coated with a tough plastic to reduce the chance of harm from breakage. Plastics contribute by providing subcomponents. Plastics have gas barrier problems. They also have strength problems that limit the use to valves and dip tubes.

There are three types of propellants used in aerosols. The majority of aerosol systems use hydrocarbon propellants. Hydrocarbons can be flammable and are often formulated with nonflammable substances such as water to reduce this flammability. Typical hydrocarbon propellants are propane, butane and pentane. They are low in cost and are satisfactory for the majority of aeorsol systems. The first propellants that were used in aerosols were the flurocarbon gases like freon. Flurocarbons are considered to be ideal propellants because they are inert, they do not affect the product, food or drugs, but they are expensive and there is the ozone problem. The use of fluorocarbons are limited and in the USA they must be approved by the federal government. Certain products that are sensitive to some gases, use high pressure compressed gases like carbon dioxide, nitrous oxide or nitrogen. The use of high pressure gases are for food items. They have a problem with pressure drop off during use because they are in a gas state only. The hydrocarbon and flurocarbon propellants are put in the container as a liquid and convert to gas as they're used, thereby maintaining a constant pressure throughout product dispension.

Throughout the years there have been many failures of products marketed in aerosols. Things such as toothpaste were never very popular because of the watered down requirement to get the product to move. Chocolate syrup developed an off flavor and also tended to be messy. There were many others aerosols that tried and failed but there were a lot that tried and were a success.

Some of the advantages of the aerosol system are controlled dispensing or exact dosage, that is required in certain medications and other products, the aerosol system tends to protect the product from contamination by not allowing anything back into the system, and it protects products from oxidation by not letting oxygen back into the package. Aerosols convert products into sprays or streams or drops or foams, and in many patterns that the consumer or packager needs.

Disadvantages are that the aerosol system provides a low product volume within the container because of the propellant requirement to be in the package, it is an expensive package and there is always the risk of failure or an explosion of the system.

There are basically four aerosol systems in use today. The system that is most used is called emulsion or two-phase system. More than 90% of all aerosol products are packaged in the two-phase system. It uses mostly hydrocarbons as propellants. Hydrocarbons go in as a liquid and then converts to a gas to maintain a constant pressure as the product is used up. There is an emulsion of the gas, the product, and liquid propellant. This system is used for paints, insecticides, deodorants, hair sprays, shaving soaps and others. Normal pressure within this package is at 40 pounds per square inch and if properly packaged the system will maintain 40 psi throughout the life of a product. A second type is immiscible or three-phase system. In the three-phase system there is a layer of gas within the can, a layer of liquid propellant either on top of the product or below it and then a product. In three-phase, gas and liquid propellant do not mix with the product. A hydrocarbon propellant floats on top of the product. A fluorocarbon propellant will sink to the bottom. Three-phase systems are not used very much and are mostly limited to powders like foot powder, bath powder, and other powders. Another type of aerosol container is the single phase or miscible. This system has high pressure gas plus the product. There is no liquid propellant in this system. Single-phase is used for foods to enhance compatibility. Carbon dioxide gas may give an acidic taste to products when used, nitrous oxide may give a sickly sweet taste. Each product/gas mixture must be studied and controlled for compatibility. Single-phase does have a low product volume because the system has to have more high pressure gas. Single-phase systems will start with a very high pressure of 100 psig or about 6.8 atmospheres and as the product is used up the pressure will decrease because of the increase of available air volume in the container. By the time the product is gone,the pressure is down to 1.7 atmosphere, which is still enough pressure to evacuate the product but the efficiency of the whole system will vary throughout the life of the aerosol system. The last system to be covered is called the plastic insert system which consists of a separate collapsible plastic bag within the can that contains the product only. The aerosol can is charged from the bottom through a charging valve with propellant. There is no contact between the propellant and the product. The propellant causes the plastic bag to collapse as the product is discharged. It is used for those items which are sensitive to any type of contact with propellants. Most packagers feel that the aerosol system has matured. There will always be some new products which are suited to aerosol systems but the cost of aerosols is a big factor along with consumer acceptance. The aerosol market will probably not be very dynamic. The big changes will be within the packaging system itself, the materials used and not the products.

Collapsible and Squeeze Tubes

One of the basic packages used in our society is the collapsible tube also called the squeeze tube. It is used as a package for viscous and semi-viscous products with a wide variety of applications and dispensers, everything from toiletries, cosmetics, toothpaste, pharmaceuticals and household products. In Europe and Asia collapsible tubes are widely used for foods.

The metal tube was invented in 1841 by a portrait artist by the name of John Rand who wanted to prevent his paints from drying out. Toothpaste was the first large volume product commercially available in a tube. Metal tubes were first made from tin, lead or aluminum. Today metal tubes are almost all made of aluminum. Aluminum is a strong, lightweight metal that will accept printing. The advantage of tube packaging is that it can be resealed to keep air out and protect the contents from humidity, oxidation, contamination and loss of flavor. Tubes close tightly and can be rolled up as the product is used and they are considered to be sanitary, portable and unbreakable. Consumers can avoid waste by squeezing out only as much or as little as they want. Some metal tubes are made from tin because of the nonreactivity or inertness of tin when in contact with the products. The high cost of tin limits use to those products that must be kept free of contamination. These products are primarily ointment medicines for the treatment of eye diseases and others. One of the chief advantages of the metal tube is no suck-back of the product which gives superior protection to the product. Metal tubes are classified as collapsible tubes because as you squeeze them they stay collapsed and can be rolled up from the bottom to force the product to the top of the tube. Collapsible metal tubes are formed from slugs by impact extrusion, on high speed automatic equipment. Once formed, they are trimmed, threaded, and annealed for softening. Tubes are then base coated, dried, printed, decorated and finally capped. If the tubes require internal linings this can also be done. Decoration is generally applied by offset lithography, the colors and design limited only by the scope of the lithographic process. Caps for metal tubes are made of either polyethylene or polypropylene. All tubes are filled through their bottoms which are left open in the manufacturing process. After filling the tube bottoms are flattened and then sealed. Metal tubes are sealed with a single fold or a double fold and then crimped. Sealants such as cement can be used for extra holding power.

Plastic tubes are used and are known as squeeze tubes. Plastic tubes were developed in the 1950's as an alternative to metal. By the 1960's they were being used extensively for products like hand creams and shampoos, and later for various other cosmetic and toiletry products like cleansing creams, facial masks and tanning lotions. Starting in the 1970's pharmaceutical and household products began to be packaged in plastic tubes. Plastic tubes are classified as squeeze tubes because, when they are squeezed to extrude the product and then released, they will almost always go back to their original tubular shape. Plastic tubes are lightweight, leak proof, durable and unbreakable. They are made from mono polymers and cannot provide complete barrier protection to some water based and oil based emulsions, some flavoring oils, volatile solvents and fragrances. Some marketing schemes prefer that the plastic tubes keep their shape and therefore are popular for certain products where it is desired that the entire shape be maintained during the life of the product. Plastic tubes can provide transparency which will show the product and the quantities of the product as it's being used. Plastic tubes are also manufactured by the extrusion process. The plastic cylinder or sleeve is manufactured first with the "head" or end molded at one end of the sleeve. Polyethylene, either low density or high density, is the plastic most used for squeeze tubes. When a stronger barrier is required as in the case of perfumes, alcohol and coal-tar products, polypropylene either high density or low density can be used. Most plastic tubes are either white or transparent but all colors are available. Offset printing is the standard method of plastic tube decoration, and in recent years the process has been perfected so that skin tones and pastel shades can be printed. Plastic tubes are also filled from the bottom and heat sealed or fused together. Some products, like hand creams, will have a large diameter cap which allows the tube to stand on its head keeping the product close to the opening of the tube for easier discharge.

Laminated tubes were developed in the late 1960's to provide a tube material that would have the advantage of plastic with barrier properties close to metal at a cost less than pure metal tubes but more than plastic tubes. From the time of their introduction, sales of this type of tube which consisted of a sophisticated web of more than seven layers of aluminum foil, paper and polyethylene film in a convoluted wind with a welded side seam, grew at a rapid rate. By 1979 annual production had reached a level of 500 million tubes with a manufacturer's value of 50 million dollars. The major reasons for the success of this laminated tube are one, that its lamination can be inexpensively modified to meet the different applications and two, its comparative costs. Laminated tubes are less expensive than metal tubes to manufacture and they are less dependent on the swings in the price of aluminum. Laminated tubes also have the advantage of having a good consumer feel that is better than pure plastic and they are able to maintain their attractiveness throughout the entire life of the contents. The first laminated tubes were used to package high volume products like toothpaste but soon many other products ranging from denture adhesives and artist paints to hair care products and pharmaceuticals began to be packaged in them. Both small volume fillers and high volume fillers can use laminated tube packaging. The heads of laminated tubes, like those of plastic, are mostly molded from polyethylene.

The laminate tube now has about 40% of the tube market while pure plastics take 30% of the tube market and metals the other 30%. The packaging industry expects in the future that there will be a continuation in the current use of collapsible or squeeze tubes and new applications. It's expected that new products will find this type of package to be an ideal container/dispenser and that other markets may use tubes like bubble gum has. New tips and closures are being developed with improved textured surfaces that offer imaginative means to have both old and new products in different ways. Better graphics and high speed automatic fabrication is expected.

Although in recent years both laminated and plastic tubes have significantly reduced the use of metal tubes, the use of metal collapsible tubes is expected to continue at its market share because metal provides a superior and necessary protection for some products. The collapsible and squeeze tube is expected to play an ever increasing role in packaging in the future.

CHAPTER 5

Packaging Closures

A packaging closure is defined as a sealing or covering device affixed to or on a container for the purpose of retaining the contents and preventing contamination thereof. The consumer expects to find the contents of a container to be as fresh when it is opened as it was on the day that it was filled and sealed. They further expect the food to be clean and tasty, the beer foamy, and the drug and cosmetics safe to use. This is true because the progress made in designing closures has had the study to meet the requirements of processors as well as consumers, and to develop a seal that's positive yet easy to open and reclose, and to produce the right closure for a specific product. Through history the search for closures parallels the people's efforts to provide requirements. The first requirement for closures was for wines. A clay cap or a wedge of wood was a natural thing to use in primitive times. Egyptians used to roll up a strip of linen to plug a jar and seal it with pitch. Another Egyptian closure was braided grass. Archaeologists rarely find closures with the containers, suggesting that they were of common natural material. Any suitable substance was used to fill an opening much like the moonshiners in this country would typically use a corn cob. When cork from Africa and Spain became available to the Greeks and Romans, they found that this compressionable material was useful as a closure. The openings of their vessels were not round, and by squeezing in a piece of cork they would get a fair fit that with pitch would make the opening air tight. Most of the urns and jars were wide mouthed and unsuited to cork, so a layer of honey or oil was poured over wine to exclude air. England's use of cork from Spain combined with the glassmaking industry were reflected in many references to corked bottles in English literature. Tapered corks were in use with the wider end protruding from the bottle for easy removal as corkscrews had not yet been invented and the wines during this era were all still wines. The long use of cork as the primary bottle closure had begun and was not to end until the 1920's. The cork closure was further established by Pierre Perignon, who was in charge of cellars in the Champagne region. He discovered how to blend white wine and retain its effervescence. His method of resisting the build up pressure was to use strong bottles and tied-on corks which allowed the safe transportation of champagne.

The United States, in the early 1800's, was a rural society with little merchandising of packages, so the incentive for developing closures existed only for a few items. The cracker barrel symbolized the era. Barrels, wooden boxes and burlap sacks lined the walls of the country store and the city general stores. Whale oil for lamps was measured out into jugs and plugged with a potato. A scoop of coffee beans was weighed and dumped in a paper bag. Syrup, cider and vinegar, which were taken from barrels to customer's bottles, were among the many items that reached the consumer without a primary package. Carrying a beer pail to the local saloon was common. Farm households ate off the land. The farmer's wife put up jams and jelly for the winter in jars covered with paper and secured with string. There was a cold cellar for vegetables. Townspeople endured the restricted diet when foods were out of season for commercial food packaging was limited. The greatest market for bottling at this time was of beverages and medicines. Whiskey was the national drink in preference to flat beer. Pocket flasks of corked whiskey were as common as change-purses. Settlers, traveling to the west in the 1820's and onward, refilled their flasks at trading posts. Most whiskey bottles were unmarked since they would be refilled from a tavern cask, but a distiller in Philadelphia named E.C. Booz had his bottles lettered and molded to the design of a log cabin in 1840. The neck was the chimney, the cork stop the smoke. Wine bottles were molded with rings for secure tying. Twine or wire was needed also to withstand internal pressure of bottles of carbonated water. Soda water was the new drink called pop. With honey or fruit flavor added, soda water gained popularity in England and America in the 1870's, taxing the facilities of bottlers and creating an urgency for quicker sealing. Cod liver oil came in bottles designed in the shape of a fish with a cork in its mouth. Stomach bitters had a label pasted across the cork and over the bottle.

The advent of metal screw closures arose from the need for home canning that went back to Appert's instructions on food preservation during the Napoleon era. The first fruit jars were tall bottles with a small or medium sized mouth that were corked or plugged with a corn cob. To make a corn cob snugger, they were wrapped in paper or a rag. When wide mouthed jars satisfied the need to insert whole peaches or tomatoes, and when corks of sufficient size were still scarce and costly, wax or lard was often used to protect the contents. One type of jar had a grooved lip affording a channel for hot wax with a pressed glass lid.

Among the patents for fruit jars with stoppers and fasteners, was the metal screw with a rubber washer that John L. Mason had invented. Mason concentrated on the quality of the glassware. Glass threads that started at the top of the neck and ended at the shoulder caused a misfit. The upper end would break in the grinding of the finish. The lower end stopped the cap from coming down to the shoulder to meet a rubber ring. Mason started a diagonal thread slightly below the top and let it vanish before reaching the shoulder. When the cap was screwed down, its rim contacted the rubber ring and achieved a strong seal. In 1858 at the age of 26 Mason was granted patents for his improved thread for blowing glass bottles. This was the birth of the Mason jar, and the start of widespread home canning.

Over the years many perfections have been added to Mason's closing system. Today we expect the effective closure to satisfy two opposing conditions. First, it must prevent contents from escaping and not allow outside substances to enter, and second, the user must be able to open and reclose the container without the use of mechanical tools. Basic screw closure operations act like the coming together of a vice. The top of the jar or the sealing surface comes in contact with the closure through the screwing operation and in between the jaws of the closure container system is a resilient material that is compressed to compensate for imperfections in both closures and the containers.

Closure Types

The common cork continues to be a popular and effective closure. The cork is the elastic tough outer tissue of the cork oak found in parts of Southern Europe, and is made up of many sided cells with a thin resinous wall. Cork cells trap air making them a compressible, impenetrable elastic and stable material. For insertion, corks are compressed and allowed to expand once inside the neck of a container resulting in a large area of contact and frictional grip while making an air tight, gas tight and moisture proof closure. Recent studies have found that plastic corks are as good in wine bottles but are not accepted by the consumers of better or more expensive wines. Tradition dictates that the old fashioned cork helps sales. One of the problems of the cork is that if it is not kept moist it will dry out and crumble in time. Corked liquid containers must be stored laying down so that the cork stays moist.

More modern type closures tend to be of the screw on type, threaded or lug, where the cap, through its engaged container threads, provides a simple means of sealing, opening and resealing. Screw type closures work for all kinds of products in both bottles and jars, and can be adapted for vacuum pack products. The lug type closure is very similar to screw on type and are used mostly for vacuum packs. They excel in ease of opening and reclosing because they only require one quarter of a turn to get the cap lugs positioned under corresponding projections on the container neck.

The crimp-on or crown closure was one of the early closures for carbonated beverage bottles. The crown closure has tremendous holding power and since the openings of bottles are small it provides excellent sealing. Crown closure removal was by use of a prying or leverage type of tool that destroyed the reclosure capability but in recent years packagers have switched to a twist-off or a tear off closure. The twist-off closure can be resealed. There are many other variations to this but it is an economical and effective closure for carbonated beverages and still is in popular use today.

Another closure is the press-on, which is limited to vacuum sealing. The advantage of the press-on closure is during filling and closing the closure is put into place by machine and pressed on so that lower parts of the skirt of the closure are forced over a retaining bead, making a good closure for high speed packaging. The press-on closure requires prying off which causes some damage during opening resulting in poor resealing.

There has been a trend in recent years to the roll-on cap. The roll-on cap is a straight sided aluminum cap that is placed on the top of a bottle that has threads. The cap threads are rolled into the skirt of the cap, after filling, providing a tamper resistant closure because the lower portion of the skirt is designed to be rolled in underneath a bead. This ring must be fractured or broken during the opening operation. Another advantage of roll-on caps is that they provide a screw-on reclosing closure.

Many containers have secondary closures that consist of many flexible materials like glassine, aluminum foils or laminates that are

60

glued to the top of a container providing two functions. One function is as a secondary barrier for protecting the product and second, they add a tamper evident capability to the closure system. Another method for secondary closures are shrink bands found commonly on luxury products and many wine bottles. They are bands of either cellulosic or polymeric material that are placed on the neck of the bottle and either heat shrunk or dry shrunk. There are also many dispensing closures. Dispensing closures are considered working closures for lots of different items where the consumer can dispense through the closure itself as a spray, drops, streams, puffs of powder and ribbons. The five basic requirements or criteria for selecting a dispensing closure are, first, that it has to perform its original purpose, second, be recognized as a dispenser, third, be easy to use, fourth, dispense at a desired rate and fifth, be reclosable or resealable. Most dispensing closures are thermoplastic, generally linerless, as they rely on the resiliency of the plastic to provide a seal.

Most screw on and some press on closures require liners, seals or gaskets that are identified as any material that can create a seal between the closure and the container. Their function is to compensate for the production tolerance or lack of precision fit between contact surfaces of the closure and the container neck, or simply to make a seal between the closure and container to prevent the product from leaking, or from environmental organisms or gases entering the container. Some of the functions expected of liners in today's packaging are, one, to prevent loss of product, two, prevent loss of vapor or gas, three, to prevent gain or loss of moisture, four, to prevent spoilage, five, to maintain sterilization, six, to prevent excess pressure, (some products build up pressure over time and need to be vented through one-way valves in the liner), seven, to maintain a partial vacuum, eight, to prevent loss of flavor or aroma, nine, to make the package tamper evident, ten, to carry identification and instructions and eleven, used for redemption. Most liners are composed of one or two types either homogeneous or heterogeneous. The homogeneous types of liners are one piece construction usually cellulose materials such as cork, felt, newsboard, polymeric, and waxes. The heterogeneous liners are composed of three layers. They have a facing material which has product contact that are usually coated papers or laminated papers with plastic. The backing material is for cushioning and are generally cellulosic or plastic foam materials. There is an inner seal which acts as a secondary liner and is made of paper, foils or plastic films.

Future work in closures will have an emphasis on child resistance and tamper evident type of closures for all goods with an increase in product usage assistance such as in dispensing and easy opening and closing types of closures. Studies show that consumers have an increasing demand for convenience in product dispensing. Consumers are saying that they choose a product over others because it has the convenience of a dispensing closure, and they are willing to pay a premium for that convenience. The number and variety of custom design closures are also growing in areas where marketers want a distinctive look. Over the years the function of the closure has changed from pure sealing and protection device to a system that has tremendous marketing appeal and influence.

CHAPTER 6

Adhesives in Packaging

Adhesives are defined as any material used to adhere one surface to another. The total adhesive production in the United States is over one billion pounds. Of the billion pounds, the packaging industry uses 25%, costing about 900 million dollars. Fifty-three percent of that 900 million dollars is used by the package maker for laminating, seaming and corrugating. Thirty-four percent of the package share is used by the packager for such things as labels, case and carton sealing. Thirteen percent is used in the manufacture of gummed and pressure sensitive tapes and coated labels. Compared to other materials, the amount of adhesives used in packaging is very small but very essential. The package system is only as good as the adhesives used to close it. Outside of packaging, adhesives are everywhere. The greatest single use for adhesives is in the manufacture of plywood. Every automobile on the road contains about two dozen different types of adhesives and a typical 747 jumbo jet has about 10,000 square feet of adhesives smeared over its surface. Even the cigarette and it's package depend on at least six different adhesives to bring the whole system together, from labelling to affixing a filter plug to the tobacco.

From a historical point of view, adhesives have been around as long or longer than recorded history. It is thought that the first adhesives used were probably natural items, such as pitch from evergreen trees, egg whites, and even human saliva was used as an adhesive. Some of the man-made adhesives started from animal parts. We know that animal glue was used over 3,000 years ago by Egyptians for securing pieces of wood together. Animal glue is made from bones from meat packing houses. Animal bones are refined and dried into a crystalline structure, then reconstituted with water for use as an adhesive. Some animal glues come from the hides and other parts of animals. Animal glue, called casein, is made from cows milk. The use of casein dates back to prehistoric times where certain parts of the milk could be removed, air dried and processed, and used as adhesives. By the 9th Century glues were routinely being made from fish, horns and cheese. Just in the past few decades have natural adhesives been improved. At the same time synthetic adhesives have come into use. Within the packaging field there is about an even split between natural and synthetic adhesives.

The first and oldest type of adhesive is called "natural" which is now made from starch-based vegetables or from protein. The more recent type is the synthetic based adhesives that are resin emulsion, latex, hot melt or solvent borne. There is a 50/50 use split in the packaging field where most of the natural adhesives are used in converting and synthetic adhesives are used mostly by the packagers for sealing. Of the natural adhesives, the starch based adhesive is most used in packaging and the paper industry uses most of that. In the United States, the starch adhesives are mostly made from corn because it is a plentiful grain available at a low price. Other countries, such as Australia, use primarily wheat. Some countries in Europe use potatoes and Latin Americans generally use tapioca. Any starch based grain is suitable for this adhesive. Starch based adhesives are water sensitive and prone to attack from microorganisms. Starch adhesives have good availability at an attractive price and, from a packager's point of view, it is a clean adhesive for using with machines. It has consistent properties—good adhesion with paper, with excellent heat resistance, and when compared to all other types of adhesives, is the least expensive. Disadvantages are that it has slow setting properties and limited water resistance that causes it to break down when moisture comes in contact with the glued surfaces. Starch adhesives are used in corrugated board manufacturing, case sealing, carton sealing, bag making, tube lining, paper laminating and labels. Starch derived adhesives can come in many viscosities which range from fluid to very heavy and stiff dextrans that are tan to brown, very tacky fluid, high in solids and excellent with labels. Borated dextrans are fluid, filmy, fast setting and very tacky with a better high humidity resistance and are commonly used for case and carton sealing, tube lining, bag seams, wrapping, and laminating. Starch adhesives also come in the form of jelly gums which are rubbery, cohesive, gummy, very tacky, good humidity resistance, excellent glass adhesion and mostly on the alkaline side for better stability. Color varies from white to amber or red brown depending upon the composition and pH. They are used mostly for automatic bottle labeling. The other natural adhesives are animal glues which are tan brown in color and supplied as dried flakes or as pre-plasticized cakes or liquids. They are considered to be very tacky but must be used at a temperature of about 140 degrees fahrenheit. As the name implies, these are products of animal parts. Animal glues (protein) are mostly used for set up boxes, tight wrapping and stripping, tube binding and film as remoistening tape. The last natural adhesive is casein which comes from milk. It's a fairly light moderate to high water resistant with good tack, usually supplied at an alkaline pH. It excels for ice proof label adhesives, for cold drinks and for foil laminations. Casein is used for returnable bottle labeling because the label is easily removed in an alkaline wash and new labels can be reapplied with casein.

The other 50% of adhesives used in packaging are the synthetic adhesives. The adhesive that is used most is called resin emulsion. Compared to the starch adhesives, the cost is two to three times greater, but it is the one that is preferred by the packaging people for sealing packages. Resin emulsion is considered to be any one of the class of solid or semi-solid organic products of natural or synthetic origin. It tends to be noncrystalline, of high molecular weight, with no definite melting point. It is mostly a derivative of polyethylene as a stable suspension of polyethylene or sometimes polyvinyl acetate particles in water. It is used in a liquid form, sometimes called the white glue because of its white color. From a use point of view it's a strong, tough, excellent adhesive for paperboard that is fast setting with little color, little taste, odor or toxicity that makes it very popular among the food packagers. Some latex adhesives are also used. This is a very expensive adhesive costing five times more than the natural starch. It consists of a small particle size suspension of latex and water. Latex can be of either natural or synthetic rubber. It is mixed with a thickener and it can have a variety of colors from white to tan. It has limited stability and machinability and high cost limits its use. Latex adhesive is used for self sealing bags, envelopes and wrappers, film bags and foil paper lamination. The next synthetic adhesive is hot melt. Hot melts are adhesives that have the most growth in packaging use at this time. It is an expensive adhesive costing somewhere between six to eight times more than the starches. The high cost is partially offset by the fact that it is all solid. A hot melt is typically a solid blend of polymer resins, plasticizers and waxes. Hot melt must be heated to liquify and use, requiring a lot of energy to get temperatures up to 250 to 400 degrees fahrenheit. It can come in many different forms that are solid at room temperature. It is purchased in the form of pellets, slats, chunks, granules or continuous ropes in bulk form so it stores and handles well. Colors can be a variety of white to brown or can be purchased in specific colors. Hot melts are considered to be the fastest setting adhesives available. Hot melts excel in high speed packaging, where they are used in case and carton sealing, bag seams, cans and bottles, labels, coatings for pressure sensitive labels, paper tape, and laminations. It has excellent water aging, minimal clean up but cannot be used with heat sensitive materials.

Solvent borne adhesives are a broad description of a wide range of polymers and modifiers dissolved in organic solvents. They are usually mobile liquids. Colors range from water-white to brown. Most are fast drying, depending on the solvent blends. Most are flammable. Most are simple nonreactive systems, but some can have properties enhanced by curing. They are commonly used in graphic arts laminations of film-to-film, paper or foil, flexible packaging, laminating film to film for pouches or bags and pressure sensitive coatings for labels and tapes. Application must be in a vented area because of toxicity to workers.

CHAPTER 7

Food and Beverage Packaging

Approximately one-half of all packaging material in the USA, both volume and value, are used for food and beverage packaging. The percentage of disposable income spent on food in the United States is about 16% and, when compared to Western Europe at 32% and some South American countries that are as high as 85%, it makes the USA's food costs very favorable. The United States is in a unique position in that it has available a wider variety and greater abundance of food than any other country in the world. This availability is attributed extensively to advances in food processing, package technology, and marketing. In the USA there are over 25,000 different food processing companies, not including the 4,000 beverage firms that are bottling carbonated and still beverages of all types. The processing companies service some 300,000 retail food stores which includes 4,400 supermarkets and 35,000 self service stores that have 90% of all sales of food and beverage for home consumed foods. A unique thing about food and beverage packaging is that it has to compete for the shopping dollar and it still must protect the product. The food package must be pleasing to the consumer and must provide directions, utility and assist in shelving or display and still be within government, federal, state and local compliance. Marketing studies indicate that the package alone will influence the first sale of an item and, if that product is accepted by the consumer, the product will generate resale, and the package will then stand as an identifier and protector. The unique thing about food is that it must be available year round in an interesting variety, irrespective of the food growing season. Foods must be presented in a way that is convenient to purchase and use, and in most instances, this means that it must be packaged. The single factor requiring food packaging is availability.

When packaging departments search for proper package materials for items, they must know four sets of facts to successfully package the product. First, they need to know about the product—the materials and the manner in which these materials can deteriorate, the size and shape of the product, the weight and density of the product, the overall weakness, breakability, bendability and integrity of the product, and how the product is to be used. Second, they must know the transportation hazards, whether the product is sensitive to heat, cold, vibration, shock and how much stacking height can the product within the package tolerate. Third, the market must be taken into consideration. They must know market lighting, the shelf life under normal storage conditions, display conditions, shelf size, location, and how high they would normally be stacked on a shelf. Fourth, they must know the forms of packages that are suitable, the machinery that's available for this packaging system, and the labor that is required to fully package and distribute the product.

Fresh fruits also need packaging. The saying about fresh fruits is that life begins at 40, meaning that above 40 degrees fahrenheit microorganisms start to thrive and cause an accelerated deterioration of fresh fruits.

Most red meats require oxygen to produce an attractive surface. If red meats are stored in an oxygen free environment, the surface of the meat will be less attractive and less likely to be selected by the consumer at a self service store. Most meats must be quickly cooled down to 50 degrees fahrenheit as soon as possible and then held at about 34 degrees fahrenheit, which is considered to be the best holding temperature. Beef and veal can be held for 21 days without suffering deterioration. Lamb can only be held for about 15 days, pork 14 days, poultry 7 to 10 days, and internal organs such as liver and kidneys, only 7 days. The quality in storage is effected by the growth of microorganisms, enzyme activity and oxidation, all of which can be slowed down by lower temperatures.

Fish products are the most susceptible to spoilage. The bacteria in seafoods are cold loving that can be active at temperatures below 40 degrees, therefore, when first harvested from the sea, these products must be packed in ice right away. Spoilage of seafoods can occur within two to three days from harvest and must be kept at near freezing temperatures.

Fruits and vegetables are unique in that they are still alive after harvest. They continue to respire by taking up oxygen and giving off carbon dioxide. They tend to be bulky, taking up a lot of space and are easily damaged by handling. Most are water based. This water can easily be lost which decreases the look and quality of the fruit and vegetable. The deterioration of fruits and vegetables starts from the time they are harvested. They are damaged both by heat and cold, and affected by oxygen, carbon dioxide, ethylene gas and other volatiles in the atmosphere. Packaging of fresh fruits and vegetables is based on a need for easy handling for distribution and the need to be chilled to reduce respiration. Fresh fruits and vegetables have two types of spoilage and deterioration. First, there is the biological spoilage which is the normal process of aging, and second, there is an anaerobic spoilage which is the due to internal reactions that will result in off-flavors, off-colors and off-textures.

Processing and packaging of some fruits and vegetables can be done. Some products can be heat processed, which destroys microorganisms like retorting canned goods, pasteurization of milks, juices and other beverages, irradiation sterilization all of which kills microorganisms. Food and beverage irradiation does not leave a toxic residue but there is public concern about its use. In the USA, irradiation is mostly used for sterilization of medical devices after they have been packaged, and very few food items are radiated. Foods can be refrigerated or frozen which slows or stops the natural processes of deterioration. The package for frozen foods serves to keep out other microorganisms and to control freezer burn. Some food products can be dried and chemically processed. Drying reduces water content to the level that microorganisms cannot live. Drying is used for pasta, freeze dried foods, beef jerky and others. Chemically processed foods are mostly salted and packaged in a vaccum package that is free of oxygen and is used for bacon, ham and others.

The food and beverage package has to be unique to the product, to its properties, to exclude gases and moisture, or in some cases, to allow penetration of some gases and moisture. In the future, food packaging will continue to receive emphasis because of its large market. The trend may be towards more microwavable food items in smaller unit size packages for easy convenience of use especially for smaller families and working families. In the future there should be an emphasis on providing more convenience and better quality of packaged foods and beverages.

CHAPTER 8

Pharmaceutical and Cosmetic Packaging

Pharmaceutical and cosmetic packaging is studied as one type of packaging because their requirements are very similar. Packaging systems, techniques and speeds are also similar.

Part One: *Pharmaceuticals*

The Federal Drug Administration defines pharmaceuticals or drugs as articles intended for use in the diagnosis, cure, medication, treatment or prevention of disease in man or other animals and includes nonfood particles intended to affect the structure of any function of the body of man or other animals. Pharmaceutical packaging manufacture is under heavy regulation by FDA and is strictly monitored. There are currently about 1,000 drug manufacturing companies in United States with at least one in every state. Some states like Pennsylvania, Indiana, New York, New Jersey and Illinois have as many as five.

Pharmaceutical packaging uses all types of packaging materials that are available, from glass sealed items such as ampules for injectable drugs to folding cartons for containing over the counter drugs. Prescription drugs require little or no sales appeal having packaging emphasis on protection and identification. Prescription drugs are commonly put into glass containers or plastic containers with good reclosure systems. Over the counter drugs must have sales appeal. Graphics play an important role and uses many folding cartons to contain the primary protection package in glass, plastics, tubes and others and must compete with all the other available non-prescription pharmaceutical products. Cost of packaging of pharmaceuticals is secondary to performance. The processing and packaging operation of pharmaceuticals is mostly short production runs, little automation and very labor intensive for quality control and inspections. Food and beverages packaging machines do a lot of the inspecting for check weighing, fill levels, missing parts and metal detection. The pharmaceutical companies rely on extensive human visual inspection.

The pharmaceutical packaging emphasis is first, on moisture control because most pharmaceuticals are hygroscopic, meaning that moisture is readily adsorbed resulting in caking, efflorescence, or inactivation of the product. Second, emphasis is on oxygen control because many drugs will oxidize resulting in reduced potency. Third, emphasis is on volatility where many of the solvents and oils have volatile organic backgrounds that may require extra headspace for one-way venting. Fourth, light protection is essential for some drugs due to degradation from exposure to ultraviolet light. Fifth—heat—since pharmaceuticals are chemical products, heat can cause accelerated toxin formation and deterioration of the product. Sixth—sterility—many items require sterility from initial packaging until use. Surgical tools, dressings, injectables, and some medications require special bacteria preprocessing in packaging rooms with specially protective packaging. Sterility can be accomplished in many ways using chemicals, heat or radiation.

Packaging for pharmaceuticals must be child proof for the protection of children. This is tested by giving potential packages to children of certain age groups. The test requires that at least 80% of these packages be intact after a specified time period. Also when testing for childproof packaging, the packages must be capable of opening by older people who might have handicaps. Emphasis is also placed on tamper-evident packaging to discourage tampering of the product. Packages containing drugs, especially over the counter drugs, are being manufactured to have either tamper-resistant or tamper-evident packaging systems. It is accepted within the packaging field that tamper proofing is not possible. Lastly, there is now an emphasis on unit-dose packaging which has reduced costs and increased accuracy in health care administration.

Part Two: *Cosmetics*

Cosmetics are defined as articles intended to be rubbed, poured, sprinkled or sprayed on, or introduced into, or otherwise applied to the human body or any part thereof for cleaning, beautifying, promoting attractiveness, on altering the appearance and also, articles intended for use as a component of any such articles except that term shall not include soap.

Soap was first used as a hair dressing by the ancient Gauls around 100 A.D. Soap had two distinct primitive uses. One was to frighten one's enemies by coating ones-self with the soap and then go around partially naked believing that the enemy would be scared of this coating. Second use was to attract the opposite sex. The second use has been more developed and proven to have more success throughout the world. In the story of "Jezebel", the book talks about using cosmetics which has connoted a long time association with loose morals that is no longer true. Eye makeup goes back to ancient Egypt where mummies in tombs have green paint under their eyes and black on their eyelids. In the 1600's, Queen Elizabeth the First, was known to take a bath once a month whether she needed it or not, but she did use perfumed water daily—up to three bottles—to overcome the lack of bathing. Her successor, James the First, never bathed, nor washed but sometimes would dabble his fingertips in a bowl of rosewater to remove exceptionally foul residues.

Cosmetics of varying degrees have been around for centuries. The manufacture of cosmetics and toiletries on an industrial scale started in France during the Napoleon era, in the early 1800's, and he himself is said to have used a pint of cologne every day. The early use of cosmetics, primarily known as toiletries, was perfumes to combat the stench of open sewers and lack of bathing.

Comparing the cosmetics industry with the food and drug industry, cosmetics has fewer government regulations. There are more smaller manufacturers of cosmetics, and the cosmetics field in the marketplace has more competition. The product protection system for cosmetics is similar to drugs because they are chemicals or mixtures of chemicals that require protection from the same things that drugs do.

The packaging of cosmetics is influenced by the user. Packages for men's cosmetics or toiletries tend to be masculine by shape and are squat or chunky, while colors are stronger, using brown, earthy tones and green for the woodsy effect. Package shapes often have masculine symbolity.

Women's cosmetics tend to have a look of elegance or luxury, hoping that they will look nice on dressing tables and the use of pastels or shades of rich colors are used. The graphics tend to picture beautiful women, suggesting that you too, the consumer, can be beautiful if you use the product. Cosmetic packages for women, from a marketing point of view, are selling hope. They almost always have continuous reformation and renaming of products to keep them up to date and to make them appear to be new and fresh. Overall they will have the look of luxury.

Baby cosmetics use simple packages with white backgrounds to suggest hygiene. They are designed to attract the baby's mother. Both cosmetic and drug packaging will tend to be similar in the future and they will be from small production companies packaging at lower speeds. Drug packaging emphasis will be on protection while cosmetic packaging emphasis will be on marketing.

CHAPTER 9

Industrial and
Military Packaging

Part One: *Industrial Packaging*

Industrial packaging is used to contain and protect materials used by industry. Industrial packages are mostly larger and heavier containers with no attempt to make them appealing. Exterior graphics are used for identification and shipping instructions and, in fact, sometimes the contents are coded to reduce the chance of pilferage. Many times the larger and heavier containers require machines for lifting and moving.

The first priority in industrial packaging protection is to maintain usability and to protect the product from damage during distribution and storage; second, is cost which is directly dependent on fragility and worth of the contents and must assume an acceptable risk of damage to keep cost within reason. Third, is convenience of use with only some reusability of the shipping container. Lastly, is appearance which is only a minor factor in design.

Package forms tend to be heavy duty shippers like multiwall paper bags for free flowing material, that can contain anything from concrete to dry chemicals. Paper is used because of low cost, the fact that it provides some circulation and is a good ultraviolet light blocker. Plastics are used by themselves or in combination with paper for moisture resistance, chemical resistance, and reduction in weight and storage space. Corrugated shipping containers are used extensively because they are lightweight, low cost, have good stacking strength for weight ratio and provide a good printing surface. Metal containers are used extensively for large liquid products, free flowing products and hardware. Metal containers are semi-portable tanks, like barrels or drums and can be as small as pails. They also can be as large as company owned railroad cars and barges. The large automobile companies purchase paints for their cars by the railroad tank car and many food companies buy sugar in bulk liquid form, shipped by railroad car. Common floor waxes are bought by the barrel and even pencils can be bought by pallet load. Portable tanks can be made from stainless steel for food and drug items. Sometimes they can be made of plastic and most all of them are reusable. Wood is used extensively for boxes, crates and blocking for high strength, wet weather storage and for stacking capability. Wood can provide good handling capability as platforms or skids for machines and large parts. Plastics are used mostly in film form for extra moisture protection. Plastic bins and barrels are used because they are nonreactive and have good stacking strength. Plastics are used as cushioning for shock or fragility protection.

Part Two: *Military Packaging*

Military packaging is very similar to industrial packaging. Specifications for military packaging are the responsibility of the Department of Defense which annually spends greater than one billion dollars on packaging for military items. Most packaging is done by suppliers of military goods according to stringent military specifications. Military specifications are very specific and very detailed because the military never knows when and where a war supporting item will be needed. Military goods must be packaged for storage under unknown conditions for long periods of time.

Products developed by the Army, Air Force and Navy fall into four categories with related packaging. The most stringent category is called Category A, which has a long term storage requirement under severe conditions and demands the most rigid type of packaging. Category B, is a medium storage length, in a sheltered condition and packaging requirement can be relaxed. Category C, is items that have immediate usage and normal packaging that would be found for commercial or industrial items are satisfactory. Category D, has no specific packaging requirements and are typically for food items and things that are going to be used soon and will not be put into storage.

Many of the items used in the military are parts for airplanes and ground vehicles that have been out of production for as long as 25 years. These parts must be purchased at the time of initial production and kept in storage for future use because producers of these parts have put away the machinery required to make the parts and they are reluctant to put the machines back into production to make one or two parts. Very few of the parts are made but they are essential for the operation of the machines and availability may require long term storage under unknown conditions. The concern of military packaging is first on preservation which requires distribution, cushioning for protection and concern in marking or labeling. Printing inks which identify the items packaged must not fade during long periods of time under all types of storage conditions of high moist heat found in tropical areas, to hot dry conditions in the desert, to arctic conditions of extremely low temperatures.

Each branch of service has special considerations that dictate packaging. The Navy requires everything to be as small as possible because of the shortage of storage space and conditions in sea going vessels and, at the same time, they must have a high corrosion resistance to salt water and high humidity. The Air Force is known to have better storage conditions but their items tend to be high value items in low supply. They must be air transportable so both weight and cube are requirements and restrictions. The Army's storage conditions are very primitive. Almost everything is put outdoors. It has a rough distribution system and must be capable of going anywhere, any way either by air, by sea, by rail or even dropped into remote positions out of aircraft. The Marine Corp requires everything to be lightweight so that they can satisfy their requirement of over-the-beach concept.

Many high value items are returned to depots for inspection and repair. The new or replacement items come in reusable containers and the out-of-commission items will be returned for repair in the same containers. The initial cost of these reusable containers is very high but the reuse capability saves money in the long run. Jet engines are shipped in metal containers that are hinged and open up like a clam shell. They can be sealed, the air can be conditioned, they can easily be loaded into airplanes, have a stacking capability built-in so they can be stacked up for storage and easily handled by common forklifts throughout the world.

A comparison of military requirements versus commercial requirements is that commercial ball bearings come in a folding carton with an inner wrapping of glassine or greaseproof paper while the military package for long term storage is often in an oil bath sealed in a metal can.

Because of the precise packaging and research requirement for military packaging, many commercial packages are fallouts from the military studies. One of the packages that was developed for the space program is the retort pouch that is now used commercially.

When food and beverage, drug, cosmetic, industrial and military packaging are compared, food and beverage has a first priority of cost, second priority of appearance, third priority of protection and their fourth priority is convenience. Pharmaceuticals and drugs have a first priority of protection, second priority of convenience, third priority of appearance and fourth priority of cost. Cosmetics have a first priority on appearance, a second priority of convenience, a third priority of cost and the fourth priority is protection. Industrial packaging has a first priority of protection, a second priority of cost, a third priority of convenience and fourth priority of appearance. Military has the first priority of protection, second priority of convenience, third priority of cost and fourth priority of appearance.

Some of the priorities are difficult to rank and may be undefendable, but when put in chart form will give a clear comparison.

CHAPTER 10

Distribution Packaging

Distribution packaging is defined as the integrated package and product handling from factory to point of sale. It includes the outer or tertiary and/or intermediate primary container which is required for efficient transportation and storage. It also includes the packaging machines and processing systems for the product. Distribution packaging can contain food items, heavy appliances or heavy industrial machinery. It can be a bundle of twelve cracker boxes, a corrugated box containing household chemicals or pallet loads of carburetors. It includes all merchandise that is to be shipped. The purpose of the distribution system is to get the right goods to the right place, at the right time, at the right quantity, at the right cost with minimum or acceptable damage. The ideal situation in distribution packaging is when the product plus the package is equal to the environmental severity. If the product and the package are greater than the environmental severity, than there is over-packaging and higher than necessary cost. If the product and the package is less than the environmental severity, then there is under-packaging and a high risk of damage in transit that could be costly. In the United States products are generally underpackaged with the producer willing to accept or assume a certain risk of damage during transit.

Distribution environmental problems are of three types. First, are physical problems which are dynamic and static and are a result of shipping and warehousing the product. Second, is climatical when weather influences the package system and moisture penetration is a big concern. Third, is a human problem which is dynamic and is the handling system. Studies over the years have definitely shown that the human handling portion of distribution causes, or is responsible for, most of the damage occurring to a product package system.

The sequence of distribution environment of a package system is product assembly and packaging where the product is subject to conveyors, vibrators and physical handling by both machines and humans. Products are then transferred to the warehouse which subjects the whole system to numerous types of handling, both machine and human. In storage at the manufacturer's warehouse, products are subject to climatical influences and the static environment of stacking as the system will allow. The products are then transferred to transportation vehicles where they again encounter physical problems as they are transported by a number of means to district or wholesale warehouses. During this process, both climatic and physical problems arise. During the transportation phase, the systems are subject to both shock and vibration. Products are then transferred to the retailer where they are subject to physical problems, both human and mechanical. Products are then displayed by the retailer and subject to human and climatical problems. Lastly, they are delivered to the final consumer and subject to human and physical problems again.

During packaging research for product protection, one must test for these known problems in sequence and try to find weaknesses in the packaging system. Studies have shown that most damage to packaging product-systems occur between local distributors and the display shelf. The conditions to be considered in the testing sequence are the atmospheric conditions anticipated, compressive loads due to stacking, forces applied by warehouse handling equipment, anticipated people handling, vibration in vehicles and impact or shock in vehicles. As production becomes more centralized, then the distribution system increases in complexity. Over the last 100 years there has been a tremendous increase in distribution cost. Comparing cost of producing products versus cost of distribution from a percentage point of view, studies show that in 1800 the manufacturer spent 70% of his dollar on producing the product to a saleable condition and 30% of the cost was for distribution or getting the product to the consumer. By 1950 cost had equalized 50% of the manufacturer's cost for product development and packaging and 50% of the cost was for distribution systems. By 1970 it was 40% for the product and 60% to get the product to the consumer. The last figures available showed that 35% of the manufacturer's costs were for product preparation and 65% for distribution. This cost comparison varies tremendously from product to product and is only a general study. Current studies show that 19% of food cost is for distribution, 27% of automobile parts are for distribution, 43% of department store items are for distribution and 49% of variety store items are for distribution.

When people are involved in handling packages, there is a significant damage problem from dropping the packages. Studies show that there is a direct linkage between the weight of the packaged object and the drop height. For instance, if the weight of the packaged product is 20 pounds or less a person will throw this object with a drop height of 42 inches. If the weight is from 21 to 50 pounds, it is usually carried by one person and will be dropped about 36 inches. If it's 51 to 250 pounds, two people will carry the package and drop it from 30 inches. If it is 250 pounds to 500 pounds, it requires light handling equipment and has a drop height of 24 inches. If it is 501 to 1,000 pounds in weight, it also requires light equipment but the drop height will be from 18 inches down. If it is greater than 1,000 pounds in weight, it requires heavy handling equipment and drop heights are reduced to around 12 inches. When this information is graphed with increasing height on the vertical scale and increasing weights on the horizontal scale the curve will slope downward from left to right and be quite straight. Packages handled by the post office receive poor handling because they tend to throw everything. The UPS system tends to have more concern for packages and handles them in a better fashion.

Studies have shown that a package, during distribution from manufacture until sale, can be handled as few as two times or as many as 50 times before reaching the consumer. Furniture and appliances are handled up to 46 times before they are in home use.

Another problem associated with handling products is acceleration, commonly measured as G's. Acceleration forces are based on the gravitational effects of falling objects and the fact that an object at rest is equal in force to one G. If an item is dropped from a height of six inches, the shock, or increase in acceleration during import may be up to 17 to 20 G's. If it is dropped from between 6 and 30 inches, it is possible to generate 25 to 60 G's on that product package system and greater than 30 inches, the acceleration can be as great as 60 G's. During the testing phase of a product package system, the "G-factor" must be identified, which is the lower limit of damage. When the shock is below the "G factor", it is unlikely that damage will occur to the product. If the acceleration turns out to be in the area that is critical because of anticipated transit shocks, then the product may have to be cushioned. Acceleration is not totally a result from dropping. Some peak accelerations in truck transportation occur at prominent highway joints which are about every 72 feet apart and can subject the load within that truck to .3 to 2.7 G forces. Bridge ramp entrances can generate 1/2 to 2 G's. Road intersections and normal wear on roads can generate .6 to 6 G's in acceleration within a vehicle. A triple railroad crossing will generate .4 to 1.7 G's. The location of the package in a vehicle is a big factor in affecting acceleration. The lower limits of the acceleration is more common and the higher limits given are exceptions that have to be anticipated in the interest of protecting the product.

Students who have ridden in the rear seats of school buses on back roads may recall going over little bridges or rough crossings where the acceleration in the back of the bus can cause floating or throwing into the air from the seat. The same thing happens in freight vehicles.

Another factor that packages are subject to is vibration. Vibration is a frequency or cycles per unit of time, and it has an amplitude associated with it which is maximum displacement. All product systems have a natural frequency which is the inherent frequency that causes a unit to vibrate. There also is a force frequency which is a produced frequency that acts on another body. These two frequencies can work against each other or with each other. If they work together and they're equal, where the force frequency is equal to the natural frequency, the condition is called resonance. Resonance is an out of control condition which can cause a maximum amount of damage. Packagers must identify the natural frequency of the product and must know what the force frequency will be during the distribution system and package the product so that the two will not be equal. This can be done by first, providing cushioning, relocating the product within the transportation mode, selecting transportation modes and may require changing the product so that the force frequencies will not cause damage. The packager must know distribution mode frequencies and some of the typical forcing frequencies from springs, tires, railroads, air and water friction and engines that are published, and are usually from two to seven hertz. Trucks under normal highway travel have suspension systems that generate a two to seven hertz, 15 to 20 hertz from the tires, and 15 to 70 hertz from the structural system. Aircraft, with products sitting on the floor and in flight, will generate two to ten hertz if they are propeller driven. If they are modern jet type transports, the frequencies are from 100 to 200 hertz. Aboard ships, if the cargo is on the deck, it is subject to 11 hertz. If it is within a bulkhead, it is subject to 100 hertz. Almost all of the frequency is from water flow along the sides of the ships. Low forced frequency of 2.2 to 7 hertz will cause the most damage because this is the natural frequency of most things. The package system must avoid a natural frequency of approximately five hertz though the use of cushioning or redesign of the product.

During package development, the packaging department must work closely with product development to avoid some of the distribution system problems by conducting shock and vibration testing to determine the product damage boundary. This is done by drop testing and vibration testing machines available in University laboratories, most large packaging companies, and at numerous independent laboratories that will do damage boundary studies for those that do not have the resources or time to do it themselves. There are four standard laboratory tests used to predict the amount of protection afforded by the package against mechanical damage. They are, the drop test and the incline impact test which produce various types and degrees of shock, the vibration test that produces various amplitudes and frequencies of vibration and the compression test that produces an imposed static load.

Packages that are subject to dropping and throwing during handling, acceleration or deceleration during transit are exposed to a wide variety of shock. Excluding such things as severe storms at sea, it is a known fact that the most severe shocks are mostly the result of human carelessness in the operation of material handling equipment and transportation vehicles, and not due to the vehicles themselves nor conditions under which packages are transported. The aforementioned tests will give valuable information if designed and conducted properly. Additional information can be obtained from shipping tests that can be used to compare different modes of transportation, and alternate packages to gain information by sending a number of test packages through regular distribution channels to various points. The testing people then go to the point of destination to make a detailed examination of the test packages and products to determine weaknesses and to gather other information. Sometimes the packages are sent on a round trip and inspected back at the point of origin. If a package product passes the laboratory test and the shipping test, it will, in all probability, receive minimum damage in the normal channels of distribution. Distribution testing is not a one time procedure. There are so many variables involved that periodic tests must be carried out to ensure a minimum of damaged merchandise. Whenever a packaging change is made or a handling procedure altered, it may signal the need for further testing.

CHAPTER 11

The Machines
of Packaging

A machine is a device that is designed to transmit or modify the
application of power for motion. It is well known that an advantage
of using machines in industry is a tremendous increase in production
speed. Packages can be formed, filled, sealed and cased, by machine
at the rate of 100 to 2,000 units per minute. Dry products can be
filled at a maximum speed of about 600 units per minute, but most
dry products are filled at less than 200 per minute. Most liquid
products are filled at a faster rate, averaging 1,000 to 1,500 packages
per minute. The big difference in filling speed is the product delivery
systems, where liquids can be measured and filled faster than dry
products. Machines also offer greater precision in measurement and
standardization. They can generate far greater strength while doing
simultaneous functions in extreme heat or cold or in toxic environ-
ments and are not subject to boredom, fatigue or psychological
problems. It is also recognized throughout the time of their use to be
cheaper than using manual labor. In most industrial situations, one
machine can do the work of 400 people. In other words, one high
speed packaging line, when taken out of service, would require more
than 400 people to do the same amount of work that the machines
can do. The purchase of machines to do packaging functions are jus-
tified by saving money through increased production, reduced labor
and increased accuracy that will gain a competitive edge on those
manufacturers that do not use machinery. It is also known that these
same machines can reduce spoilage by increasing quality control and
accuracy.

Some of the things that are desirable when installing packaging
machinery are, that packaging lines run from left to right as the
operators face the control panels. Machines that run from right to left

can be purchased, but they must be specifically ordered and will cost more money than standard machines. Machine lines should be in a straight line to reduce problems of spoilage caused by jamming and shuffling of packages when they are required to turn corners.

Packaging machines were not available prior to 1900. Packages were hand filled by being dipped, ladled or siphoned. Early paper packages were made in cone shapes from sheets, and this system is still used today in small European shops. Around 1900 everything was available for making semi-automatic machines, the technology, the machines themselves and the power. Packaging operations that were being automated were package setup, product metering, package closing, package unitizing and cartoning. In 1894 the first automatic boxmaking machine became available. In 1895 "Pneumatic Scale", which is a large manufacturer and supplier of packaging machinery, made the first automatic weighing and filling machine. In 1900 Justin Bemis made a machine that could make 5,000 paper bags each day. In 1903 Libbey-Owens made the glass bottle making machine. In 1904 the first tea bags came upon the market and were accepted by the consumer. In 1906 the first successful automatic packaging line was installed in Detroit for packaging soap powder. This resulted in significant cost reduction for the customers and forced other soap powder manufacturers to invest in automatic machines. Also in 1906 was the first cut, form, and twist wrap for candy kisses. Still, there wasn't a big demand of individual packages as we know them in today's supermarkets until after World War I, when the resistance in food packaging began to change from bulk to unit packaging. Now, companies or grocery stores could not survive without unit packaging in a self service basis. In 1932 the vertical form, fill and seal machine was put on the market. It made the famous pillow package which is common today. Vacuum packs were developed in 1947 which brought the gas flushed package that allowed an increase in shelf life of oxygen sensitive products.

Prior to World War II most machines were designed for a specific requirement. For instance, Kellogg's would have a specific require-ment for a particular cereal and Lever Brothers would have a specific requirement for a particular soap powder. After World War II, machine manufacturers and packagers started the first serious effort to fully mechanize and integrate entire package lines and standardize these machines so that there was greater ease of maintenance throughout the country.

Today, manufacturers could not compete without automatic machinery unless they produced a unique small volume item that does not have competition, or the product requires extensive human interaction for quality control like certain pharmaceutical products. Machines of today can make bags at the rate of 2,000 per minute and the package filling machines are limited only by the product itself. In the future, packaging machines will continue to have more automation and integration of machines with computer controls. There will also be an establishment of a role for robots in areas where robots can perform better than humans. There are about 800 companies in the United States that make a large variety of packaging machines. Most of these companies make a specific machine that can be integrated into another company's machines. There are only a few companies that will manufacture complete lines of machinery from start to finish.

A typical dry filling line used for packaging cereals or candies that use a paperboard carton with a flexible liner would be like the following. At the beginning of the line, there would be a liner former and a carton former which would use roll stock to form a liner and insert that formal liner into a carton which would have had the bottom and side seams glued. The carton/liner would proceed through a filler which would weigh and dump the product into the formed package. It would proceed to a check-weigher which would double check the weight of the contents, and either accept or reject that package. Rejected packages would be removed from the line for destruction or recycling. From the check weigher, the package enters a metal detector which again would accept or reject contaminated packages. Modern metal detectors can detect all types of metal. The package then is sealed by sealing the liner through a variety of methods and the carton flaps are sealed with an adhesive. The sealed package is code marked for product control for the consumer and the manufacturer. Consumer information is printed to give a shelf life use date. The manufacturer's information is printed in code to record the date of filling, the batch of the product, the packaging line, the shift, the plant number and other information needed for control. The package then enters a case packer which automatically sequences and patterns the unit packages for the shipping case which automatically erects, fills and seals the shipper. The shipping case is code-dated for warehouse control with similar information that was printed on the unit or primary package. The filled and sealed shipping case can be check weighed to show missing packages. It next goes to an automatic palletizer which will pattern, stack and secure a pallet load of filled cases, ready for the distribution system.

Packaging machines all work together. The reliability of these machines must be high because two machines working together are only as good as the combined efficiency of the two machines. For instance, if Machine A is 90% efficient and Machine B which is fed by Machine A is 90% efficient, the overall efficiency will be 81% and this will continue down through the complete line. Machine A, set to run 500 packages per minute may, at the end of the day, have a net production of 400 packages per minute. Any machine that stops for maintenance will stop the whole line. Machines do require maintenance, do require servicing and do require packaging stock, all of which can cause line stoppage and production delays.

Liquid filling lines are a little different. Liquid lines start with an unscrambler which unscrambles pre-formed plastic and metal containers. Glass containers are shipped on pallets or in boxes and require a pallet unloader or a box unloader. Some plastic containers are blow molded by the packager just prior to filling. Most liquid containers must be cleaned by washing, sterilization, or an air blow. After cleaning, the package is filled with a measured volume of the product, then immediately closed by a capping or lidding machine. Closures may include fitments which aid in the dispensing of the product and can be pumps, shaker heads and others. There are many types of sealing systems used for liquid containers. The product/package is next checked for fill level by a machine that measures the level of fill, and it will reject under/over fills or accept properly filled containers. A foreign object detector checks each container for all types of foreign objects, metallic or nonmetallic, and will reject packages that check positive. Labels are applied to the container and the package is code marked with the same types of information that is done on the dry-fill lines with consumer and manufacturer's information. The next machine is the case packer and sealer that erects the shipping case and inserts the liquid containers in a variety of patterns. Glass containers are dropped in from above to accommodate partitions that are put in the shipping case to avoid glass to glass contact. Metal cans and some plastics can be pushed in the shipping case from the side. The filled cases are code marked for warehouse control and shipping instructions, and can be check-weighed for missing liquid containers. The cases proceed to the automatic palletizer which patterns and loads the pallets. Loaded and secured pallets then enter the distribution system.

CHAPTER 12

Packaging Laws
and Regulations

The packaging industry, like any industry that deals with the public, must be regulated by laws and regulations to protect the consumer from a safety point of view and a quality and quantity point of view. Laws are written by legislatures as rules of conduct or actions that are prescribed or formally recognized as binding or enforced by a controlling authority. People that are guilty for bypassing laws are responsible for these actions and can be fined or jailed by the courts. Regulations are authoritative rules dealing with details or procedures, usually written by various agencies, and are enforced through licensing and/or fines.

From a historical point of view, packaging laws were enforced as early as the eighth century where Arabian merchants were certified by a bureau of standards. This bureau would certify these merchants by marking their measuring cups and bowls with an official seal, much like scales and gas pumps are measured and certified today by various departments of governments. The problem with these early laws and certification was that many of the officials could easily be influenced or bought off. In medieval England bread makers were controlled by laws on the sizes of bread made. The bakers were identified on every loaf of bread that they baked by specific marks that were put into the baking pans and could be used to enforce the law. In London in 1373, laws required a specific mark on bottles and other vessels made of other materials, such as leather, to ensure a full measure. Up to 1800 it was common to buy bread that was contaminated with alum to extend the bread flour which was more expensive. Alum is potassium aluminum sulphate and can cause vomiting. It is an astringent which causes puckering or drawing together of soft organic tissues. Beer was altered with various narcotics to enhance sales. Tea leaves were commonly colored with a copper salt to enhance the green color of the tea leaves.

In 1819, druggists in London proposed regulations for warning labels on drug containers that were likely to produce serious mischief if improperly used. They required that these drugs could only be sold to those old enough and experienced enough to handle them. It was common to put clay, chalk, or sand in salt and sugar to increase the quantity available for sale. It was common to put water in alcohol as sales increased to assure availability of the product to the consumer. This alcohol problem by itself led to the proof system which was established on the fact that a mixture of a minimum of 50% alcohol in water would support combustion. It was accepted that if a sample of alcohol would support combustion, it was considered to be 100% proof, that it was a good product and anything less than 50% was not acceptable. Early inspection systems were corrupt. That brought on the requirement for an inspector's name to be put on all certifying seals on a product and gave some control of quality to these products. Throughout history, solutions have been based on laws passed in reaction to crises, rather than on science. For instance, when many children died from dyptheria many years ago, the Biologics Control Act of 1902 was written. In 1906 many soldiers died after consuming canned bully beef that was preserved in formaldehyde, resulting in the food and drug law. In 1938 many people died after taking an elixir of sulfanilimide. This resulted in the establishment of the Federal Food, Drug and Cosmetic Act. In 1958 there was a large group of physicians from all over the world who met in Italy to form the Food Additives Act to protect the consumer.

In 1962 there was a fertility drug which caused birth defects in Western Europe. The effect was that a new drug act was established. By 1970 there were over 2,000 children who were accidentally poisoned in the United States from consuming household chemicals that were not properly marked. This brought about the Poison Prevention Act. This was followed by studies that reported many allergic reactions to cosmetics by consumers and this precipitated the Cosmetic Labelling Act. There were many questionable medical practices using radiological cures for cancer and other problems that were useless. This brought on the medical device amendment to protect consumers from useless medical treatment. Recent tampering of over-the-counter drugs in this country has resulted in new federal laws for the sale of certain drugs to the consumer. Until the mid 1950's, the main purpose of legislation was the control and standardization of products.

Packaging was involved mostly by certain labelling requirements that stated what had to be on the label, such as the brand name, generic name, contents, ingredients and the manufacturer or controlling agency of that product. The departure from this historic relationship was created by the Food Additives Amendment Act which for the first time shifted, from government to industry, the burden of proof as to whether an additive was safe before the product was put on the market. This law also defined an additive as any substance that can reasonably be expected, either directly or indirectly, to become a part of the product. Therefore, contact between a food product, the equipment in which it is processed and the package in which it is contained became significant because any chemical migration from either source to food is, by definition, an additive. This law also contains the clause which states that no additive shall be deemed safe if it is found to induce cancer when ingested by man or animal.

Packaging laws have had many changes over the years and will continue to have changes as chemical analysis is improved and technical progress on how chemicals react in the body is better defined. The big move toward greater control of packaging started in the late fifties and early sixties by the Fair Packaging and Labelling Act which gave the Department of Commerce authority to determine improper proliferation of package size, weights, measures or quantities of packaged items and to specify how packaged goods should be identified to satisfy this requirement. There are many other special packaging regulations such as the labelling of all alcoholic beverages under the control of the Federal Department of Treasury and Bureau of Alcohol, Tobacco and Firearms. Demand by consumers for more definite statements on food and cosmetics as to the chemicals and quantity of product ingredients, in the manner similar to that already used in drug labelling, has resulted in ingredient labelling, regulations requiring the listing of ingredients in order of predominance. Any food that makes a health claim must list the nutritive value of the product. There is also a great variation in local laws.

Some states require deposit and return on beverage containers while many states do not. There are other local laws which manufacturers and packagers have to comply with. Some of the principal federal agencies that regulate packaging are: the Federal Trade Commission which is active in deceptive packaging and labelling as it relates to unfair trade practices, the Environmental Protection Agency which is responsible for packaging of insecticides, pesticides, rodenticides and air, land and water standards, the Department of Justice which, under the Drug Enforcement Administration, is responsible for controlled substances such as narcotics, the Department of Transportation which is responsible for the distribution of hazardous materials by air, land, and water, the Consumer Product Safety Commission which is responsible for hazardous household products, flammable fabrics and children's toys and articles, the Department of Labor under the subdepartment of Occupational Safety and Health Administration for packaging machinery and plant operations, the Department of Agriculture which regulates fresh and processed meats and poultry, the Department of Treasury which regulates alcohol, tobacco, firearms, explosives and imported goods, the Postal Service which has control of all mailable goods, and the Food and Drug Administration which is responsible for food, drugs, cosmetics, medical devices, and public health. Common carriers, trucks and railroads, have rules which outline how certain things must be packaged before they will be responsible for shipping of these goods. Anybody who ships goods by a common carrier must be aware of these rules and regulations and abide by their direction. The numerous laws and regulations governing packaging and its various aspects are extremely immense and somewhat overwhelming. The long run effect of this strict regulation could be detrimental to the future of packaging research and development. However, changes are being made and will continue to be made. The packaging laws do protect and aid the consumer and without them the marketability of products would be out of control.

CHAPTER 13

Graphics In Packaging

Graphics is defined as the design and decoration of the exterior surface of a package plus the use of related equipment. Packaging graphics include illustrations, symbols, color and words. Good packaging graphics requires the skills of the behavioral scientist who knows the psychological aspects of the consumer, the artist who can design and lay out the overall looks of the package, the ink maker who can match the colors under all lighting conditions and make the ink durable under the conditions of which it's going to be subject, the material supplier who provides the paper, plastics, metals and other surfaces for printing and the printer who puts it all together to establish the final product. Please note that no mention is made of a packaging expert, as the packager is more concerned with the functional aspects of the package.

Good graphics that are required in today's competitive society are very expensive, costing as much as 35% of the overall package costs. Graphics are expected to identify products, describe products, picture products, sell the products and tie the product into any advertising that is conducted in newspapers, magazines, and on television.

The most outstanding feature on a package is color and the combination of colors because they will attract the consumer's attention. Also, colors provide instant recognition and recall of products. Studies have shown that simple colors have best sales stimulation. Odd and exotic colors are not effective in sales stimulation. Color preference will vary constantly, year to year, season to season, within economic cycles. For instance, recessions tend to favor bright and cheery colors while good times are reflected best by conservative colors. Also the sex of the consumer is influenced by the colors. For years the most popular color for sales appeal for women has been red, while men are mostly stimulated by blues, very small children are stimulated by red and older children tend to be influenced by yellow. Middle aged adults seem to prefer subdued colors while older people turn to gay colors. Colors are also influenced by the marketplace lighting, the locations and from one country to another. People working in packaging color combinations have to be aware of psychological aspects of the consumer both in this country and, if exporting, must be aware of color combinations in other cultures.

Recent studies have shown that there are over 15,000 items in an average supermarket. The typical shopper of today spends about 15 minutes in the supermarket, which makes the importance of color and sales appeal of the package even more significant from lack of time for selection of merchandise. One item is competing with 15,000 items for visual stimulation and has, on the average of, .06 seconds exposure.

CHAPTER 14

Careers In Packaging

The industrial world now recognizes that packaging is a legitimate profession. The packaging profession has evolved from a necessary cost of marketing goods to an essential part of successful consumer sales and an improved protective container for damage prone products.

There are several universities that now offer four-year degree programs leading to a Bachelor of Science degree in packaging. (See Chapter 15). There are also a number of trade schools that offer two-year programs that train packaging technicians and packaging machinery mechanics, and numerous universities that offer packaging courses in conjunction with other major fields of study.

Bachelor of Science degree requirements vary from school to school but most require a combination of science, math, engineering and business to supplement and strengthen a core of packaging courses. The packaging core requires about 20 semester hours or 30 term hours of study that concentrate on the packaging discipline and cover packaging topics as; packaging materials from manufacture, conversion, uses and limits; packaging dynamics from causes, limits and prevention; packaging machinery from types, speeds, line layout, machine integration and production control; and a cap-course of case studies and applied problems. There are also numerous elective courses offered to packaging students such as, food packaging, pharmaceutical packaging, machinery components, robotics in packaging, computers in packaging, economics of packaging, solid waste and recycling, packaging distribution and logistics, laws and regulations in packaging and other special topics.

Most packaging programs offer three to six month hands-on internships with the packaging industry and there are overseas study programs in England and Sweden. A few universities offer Master of

Science degrees in packaging with graduate level packaging courses and a thesis research requirement.

The demand for trained packaging professionals in industry has been strong throughout the years with starting salaries that average well above most other University degree programs.

The future of the packaging profession looks strong. The ever changing demographics of the United States consumer will require changes in current packaging and the development of new packaging for current and new products to satisfy market needs. Anticipated strong regulation of solid waste will require continuing research to improve recycling of materials and the development of biodegradable materials. People that are interested in studying professional packaging can contact the packaging department of a near-by industry for advice on training or contact one of the universities or colleges listed in Chapter 15.

Professional Packaging Schools and Universities

Arizona State University
Department of Design Services
Tempe, AZ 85287

California Institute of the Arts
Design School
24700 McBean Pkwy.
Valencia, CA 91255

California Polytechnic State Univ.
Graphic Communications Dept.
San Luis Obispo, CA 93407

Carnegie Melon University
Director of Design
Schenley Park
Pittsburgh, PA 15213

Chapman College
Food Service and Nutrition Dept.
333 N. Glassell Avenue
Orange, CA 92666

Clemson University
College of Agricultural Studies
Dept. of Food Science
223 P&A Building
Clemson, SC 92631

Cranbrook Academy of Art
Box 801
500 Lone Pine Road
Bloomfield Hills, MI 48013

Fashion Institute of Technology
Packaging Design
227 West 27 St.
New York, NY 10001

Guilford Technical Institute
Packaging Department
Jamestown, NC 27282

Hennepin Technical Center
South Campus
9200 Flying Cloud Drive
Eden Prairie, MN 55344

Indiana State University
School of Technology
Terre Haute, IN 47809

Inver Hills Community College
Packaging Technology
8445 College Trail
Inver Grove Heights, MN 55075

Joint Military Packaging Training Center
Aberdeen Proving Ground, MD 21005

Kent State University
School of Art
Kent, OH 44242

Michigan State University
School of Packaging
East Lansing, MI 48824

Cornell University
Department of Food Science
Stocking Hall
New York State College of Agriculture and Life Sciences
Ithaca, NY 14853

New York University
Center for Graphic Arts Management and Technology
80 Washington Square East
Rm. 53
New York, NY 10003

North Carolina State University—Raleigh
Dept. of Food Science
Box 5992
Raleigh, NC 27650

Parsons School of Design
Industrial and Packaging Dept.
55 Canal Street
Providence, RI 02900

Pratt Institute
Graduate Design Programs
200 Willoughby Avenue
Brooklyn, NY 11205

Rhode Island School of Design
Industrial and Packaging Dept.
55 Canal Street
Providence, RI 02900

Rochester Insitute of Technology
Dept. of Packaging Science
One Lomb Memorial Drive
Rochester, NY 14623

Rutgers University
Packaging Science and Engineering
P.O. Box 909
Piscataway, NJ 08854

San Jose University
Division of Technology
San Jose, CA 95192

St. Francis College
Dept. of Management
180 Remsar Street
Brooklyn, NY 11201

Sinclair Community College
Engineering Packaging Technology
444 W. Third Street
Dayton, OH 45402

Spring Garden College
Industrial/Technical Division
102 East Mermaid Lane
Chestnut Hill, PA 19118

Texas A&M University
Dept. of Mechanical Engineering
College Station, TX 77893

University of Akron
Dept. of Art
Akron, OH 44325

University of Cincinnati
College of Design, Architecture and Art
Cincinnati, OH 45221

University of Illinois
905 South Goodwin Ave.
Urbana, IL 61801

University of Lowell
Plastics Seminars
Continuing Education
40 Karen Road
Waban, MA 02168

University of Minnesota
Dept. of Food Science
1334 Eckles Avenue
St. Paul, MN 55108

University of Missouri-Rolla
School of Engineering
301 Harris
Rolla, MO 65401

University of Southern California
School of Engineering
Los Angeles, CA 90007

University of Wisconsin-Extension
Dept. of Engineering
432 N. Lake Street
Madison, WI 53706

University of Wisconsin-Stout
School of Industry and Technology
Menominee, WI 54751

Widener College
Science Group
Chester, PA 19013

Wisconsin Indianhead Technical College
New Richmond Campus
1019 South Knowles Ave.
New Richmond, WI 54017

CHAPTER 16

Glossary

Aerosol: A gas-tight container fitted with a pushbutton dispensing valve and pressurized with a propellant gas which forces a product from its container when the dispensing valve is opened. Three types of propellants are employed: fluorocarbons, hydrocarbons, and inert gases.

Ampule: A small thin-walled glass or plastic container with a narrow stem or neck sealed by fushion after filling. Commonly used for the portion packaging of sensitive drugs or expensive cosmetics and food ingredients. Opened by breaking the stem.

Barrel: A bilged cylindrical container of greater length than breadth, having two flat ends or heads of equal diameter.

Blister Packaging: A packaging method in which the product is secured in a rigid dome or bubble of plastic commonly mounted on a sheet of paperboard.

Carboys: Large containers shaped like bottles and made of heavy glass as well as clay, earthenware, metal, or plastics and designed to be encased, usually in wooden outer crates. Uses include shipment of industrial liquid chemicals and potable water.

Carton: Folding boxes generally made from paperboard, for merchandising consumer quantities of products.

Case: A non-specific term for a shipping container.

Cellophane: A transparent film made of viscose (regenerated cellulose). Once dominant in visible packaging but now largely replaced by plastics.

Check Weigh: Placing one consumer package on a scale to make sure that it contains as much of the product as the label states.

Chipboard: Recycled paperboard often covered with a thin layer of bleached virgin fibre and/or a clay coating which facilitates printing. Also called "newsback" or "patent-coated".

Closure: A sealing or covering device affixed to or on a container for the purpose of retaining the contents and preventing contamination thereof.

Corrugated Paperboard: or simply **Corrugated:** A sandwich construction of layers of paperboard where a fluted medium is adhered to a flat sheet of linerboard of several standardized weights on one or both sides (single or double-faced). Second or third layers of single face board and liner can be added for extra strength.

Distribution System: A system of package and product handling encompassing manufacture, shipment, storage and marketing. A distribution channel is generally one way in nature, designed to carry goods from point of manufacture to point of consumption, but not in the reverse direction.

Fibre or Fiber: The threadlike units of vegetable growth that form the basic structural components of paper or synthetic filaments used in similar sheet materials. "Fibre" also refers to the finished products, e.g., thread and paper. Wood fibers (pulp) are the most desirable source of paper and paperboard. "Secondary" fibre, reclaimed from consumer and industrial waste, has recently surpassed in quantity the general use of "new" fibre.

Fluorocarbons: A class of gaseous compounds produced by substituting fluorine for hydrogen in hydrocarbons. Use as an aerosol propellant has recently been banned because of a suspected impact on the ozone layer of the stratosphere.

Flute: A rib or corrugation on a surface; one of the undulations of a corrugated material.

Friction Plug: A closure held in place by friction. Such fitments are usually used to control flow from bottles.

Gas Packing: Packaging in a gas-tight container in which any air has been replaced by a gas that contains no free oxygen.

Hydrocarbons: A class of compounds containing hydrogen and carbon such as butane and propane, which are used in packaging as aerosol propellants.

Impact Extrusion: A method of shaping metal or other malleable materials by charging a die with a pellet or disc of the material and, by impact, forcing it to conform to the shape of the die. Collapsible tubes and some cans are formed by this process.

Kraft Paper and Paperboard: Paper and paperboard derived from chemical wood pulp made by digesting wood chips in an alkaline solution consisting chiefly of caustic soda and sodium sulfate. The strongest of wood pulp products, kraft paper in bleached or un-bleached forms is used extensively in flexible wraps, folding cartons, multi-wall bags and as the liner or outer walls in corrugated board.

Laminate: A product made by bonding together two or more layers of material or materials.

Mandrel: A cylindrical or rectangular core around which paperboard, aluminum, plastic film, or other materials can be wound to shape container tubes that are then fabricated into drums and canisters.

Mobile Tank: A tank used for shipping unit loads of liquid chemicals and often fitted with interface pipes and valves for direct hookup to customer manufacturing systems.

Multiwall Bag or Sack: A form of flexible packaging consisting of from two to six plies of paper, asphalt, plastic films, or metal foil.

Ovenable: Capable of being placed in an oven (microwave or conventional) with food and will serve as the cooking utensil.

Overpackaging: A condition in which the packaging of a given product is viewed as having exceeded the requirements of product containment and/or protection.

Pallet: A platform of wood, paperboard, plastic or metal on which goods are placed for storage or transportation as a unit.

Palletize: To place goods on a pallet for shipping.

Paperboard: Distinguished from other kinds of paper by greater basis weight, thickness, and rigidity. Paperboard refers to sheets 0.012 of an inch (12 points) or more in thickness. Incorrectly termed "cardboard".

Polyethylene: A plastic polymer of ethylene extensively used in packaging. It has strength and low moisture permeability and good resistance to acids, alkalis, and inorganic chemicals, and is formed into films, coatings and semi-rigid molded containers.

Pouch Package: A flexible container made from a film or combination of films, papers and foils.

Pulp: The basic product in papermaking, a suspension of fibres in water. Molded pulp packaging components are made by pressing out the water in a mold. Commonly used for egg and fruit and vegetable cartons and for cushioning of industrial items.

Recycling: The reuse of materials after consumption. In contrast to resource recovery, recycling refers to a second cycle of the material for its original purpose.

Retort Pouch: A flexible container of plastic films or film, paper and foil combinations that can withstand the heat of sterilization used to preserve, food products. Sometimes termed a "flexible can". See also POUCH

Shrink Film: A plastic film oriented by stretching during manufacture which attempts to retain its original dimensions on reheating. Used as a low-cost wrap for individual products and unit pallet loads.

Single-web Rolls: In flexible packaging, monolithic structures of a single material such as paper, plastic, or foil in either coated or uncoated forms.

Skin Packaging: A packaging method in which a thin film is drawn down over a product mounted on a sheet of coated or uncoated paperboard to which the plastic adheres.

Thermoform: A plastic sheet that has been heated and then shaped by forcing it into or over a mold. Plastic thermoforms enable quick package changes and shapes not obtainable with other materials, an advantage accounting for the rapid development of kits and packages with complex and varied shapes.

Unitized Pallet Load: A pallet load in which all containers and individual items have been consolidated for unit handling or transportation. Also termed unit-load palletization.

Unit Portion Packaging: A method of convenience packaging in which the product is pre-measured in quantities of anticipated use.

Universal Product Code: A printed code on containers using symbolic identifying marks for the purpose of providing information on the product, primarily for inventory control and retail pricing.

Vacuum Packaging: Packaging in containers, rigid or flexible, from which substantially all air has been removed prior to final sealing.

Vial: A tiny, thin-walled glass container often used in the portion packaging of sensitive drugs. A vial for injectables is closed with a complex rubber-stopper-and-metal seal that permits direct withdrawal of the contents into a syringe.

Web: The roll of paper, foil, film or other flexible material as it moves through the machine in the process of being converted.